# The Seventh Day

# The Seventh Day

*Ed Dirazar*

Writers Club Press

San Jose  New York  Lincoln  Shanghai

The Seventh Day



Writers Club Press
an imprint of iUniverse.com, Inc.

For information address:
iUniverse.com, Inc.
5220 S 16th, Ste. 200
Lincoln, NE 68512
www.iuniverse.com

ISBN: 0-595-16472-2

Printed in the United States of America

*Dedicated to the Creator of all physical laws existent in the universe. Also to the few left true geologist, those who still carry the geological pick or a pair of old field boots in the trunk of the car. An endangered species of professionals in many cases forced to evolve into an environmental or engineering accessory, or even worst into an attachment to a computer on a desk.*

# Contents


Preface .................................................................ix

Acknowledgements .......................................xv

Chapter One A man of this World ...............1

Chapter Two The Beginning ........................7

Chapter Three Those Were The Days… .........23

Chapter Four The Moon Of The Fourth Day ...........29

Chapter Five Origin Of The Moon ..............38

Chapter Six The Pacific Womb ...................48

Chapter Seven A Can Full of Stars .............67

Chapter Eight Moving Through Space .........84

Chapter Nine The Expanding Universe ........90

Chapter Ten World Evolution .....................95

Chapter Eleven True And Relative Time .......106

Bibliography ............................................119

# Preface

The idea of writing this book started in 1969, on the same day when Neil Armstrong and Ewin Aldrin walked on the Moon to begin the era of spacial exploration away from our Earth.

At that time I was a young geologist doing exploration work for base minerals in the bush of southern Zimbabwe.

That day, after finishing my shower, or better said after the twenty liter drum filled with warm water went dry, I approached the group of workers sitting around the fire, to share with them the great news I just heard on the radio.

I stood watching to the upcoming full-Moon in the colorful sky of the African sunset and taking advantage that everybody turned respectfully silent on my approach I said something like, "Look at the Moon…It's beautiful is isn't it?…It seems unbelievable that there are two men there right now…"

I still remember seeing the white of their wide opened eyes in the darkening evening, looking at me and wandering what I was up to.

So I decided to break the news,…"Oh, I am not joking now…I just heard the news on the radio, the Americans have landed in the Moon and one man has been walking around there…"

There was a long and tense period of complete silence as I saw the white of their eyes getting larger in disbelief until somebody started laughing and then everybody joined him loudly.

The *Madala* (old wise man), senior of all the group, waited till everybody was again silent to tell me the "hard facts",…"Excuse me *baas*, with all respect let me tell you…We cannot believe your story about a man being on the Moon because we know for certain that no man can go to the Moon…That is Mungu territory…Mungu has a guardian there who is a son of the devil and his job is to eat people alive…"

After this announcement all voices came alive in their native language obviously telling each other different stories about Mungu and shortly everybody forgot about my presence.

Hiding my frustration I decided to retire to my tent to continue reading my old fiction novel.

On the following morning I got the local newspaper on a nearby hotel, showing photographs of the astronauts in the front page and telling a long story about the preparation and achievements on the trip made by Apollo 11 Mission.

I thought that was going to be enough to prove my point and get back these people's credibility but, to my surprise, when I displayed the paper on the camp table, I only got more laughter and disbelief and even sensed that some of the younger men were getting angry as I heard the word "nonsense" pronounced few times around.

Then I realized there was some old tradition broken and I had the feeling that they were blaming me for trying to force a change in their beliefs…Since that moment we all tacitly agreed never mention this subject again.

Now I am sure that all that is past and forgotten and these people and their descendants grew up into the knowledge that modern technology is pushing us to learn and accept new ideas that eventually will change our way of living.

As for myself, the learning was that even if you have the knowledge of some truth, there are times when it is very difficult to get other people to believe it. That was the beginning that opened my chain of thoughts and brought me to think about our ancestors in biblical times.

How could I have explained to them what we all know now about our universe and the forces that formed and reshaped our Earth...? ...Well, surely the best way is written in the bible.

After almost two thousand years of learning, even if we have a much better idea about the universe, our world and our own creation, we still continue discussing the meaning of the bible, a book that not man have been able to prove wrong and still have vital data remained to be interpreted by experts.

*...By the seventh day God had finished the work he had been doing; so on the seventh day he rested from all his work.*

*And God blessed the seventh day and made it holy, because on it he rested from all the work of creating that he had done...*

...That is the end of the story on creation according to the bible, all God's colossal tasks were completed. All these detailed wonders of nature He planned, changed, erased and improved as the biggest artist, for the benefit of His most selected creation...man.

And then, he decided to rest, allowing men to take over the control of our planet, expecting that after some time, and getting good opportunities, his children of creation would grow up to become perfect human beings, gaining a place very close to himself.

Now we all know things did not work that well as we still are in the period of learning and might even be getting further away from perfection.

Today scientists are trying to find all clues left by God on each step of his creationist work and this search goes from the smallest particles of an atom through all existing matter on Earth and the solar system with projections to the whole universe but, admittedly, we are still too far for getting close to the real meaning of the whole thing.

The book of Genesis show us how carefully God was planning every phase of his creation, going step by step building and improving everything until he saw it was good and when satisfied, he was pleased to announce it at the end of each day that marked each stage of creation.

In this book we used geo-chronological data trying to fit the main geological events on Earth formation, within each day as told in the bible, finding it could be a very close correlation.

There are still many gaps to be filled by applying new scientific evidence to this work and admittedly, they might be some flaws in our data assessment but, considering the vastness of the subject, allowances should be permitted on this attempt to unify some major events of our universe.

An example of this is our study on the origin of the Moon. Our analysis of all the geological evidence on Earth would point to an age of formation at the end of Jurassic or Cretaceous, between 90 and 140 million years ago.

This is a new concept we try to introduce here but, sampling of impact craters on the Moon are true dated between 3.8 to 3.1 billion years ago, meaning that if the interpretation is correct this event might happened in the early Precambrian and this is our second option which is also a new concept compared with the general belief that the origin of the Moon took place at a very early stage when the Earth itself was in formation.

There is a need of more data interpretation on the rock samples taken on the Moon in order to determine their true age, and also we need more knowledge on metamorphic processes in different conditions away from our Earth to build reliability on the sampling results and find exactly the sequential dates of origin.

The ideas put here on the creation and growth of the universe from the Big Bang to the end at the edge of the universe are only philosophical approaches that clearly recquires to be followed with further investigations.

Science is continuously trying to find an answer to all these questions and moreover is trying to understand the meaning of the creation of the Universe. Lately we have been making great progress in learning about all the factors involved in this process. There are many new theories continuously coming out from different schools of cosmology but, the more we find, the closer we get to conclude that everything started in a single event when, from nothing, we had the beginning of space-time, light, matter and evolution of everything into life. All of it, our God's creation.

# Acknowledgements

*I am grateful to my old friend Dr.Zvone Grobenski who helped me with valuable comments after reading the first draft of this book. I do not know if he still carries a pair of field boots in his car but I know his "geological" heart is still in the Ruwnezori Mountains in East Africa.*

# Chapter One

## A man of this World

God created an almost perfect universe with and almost perfect Earth surrounded by other planets where he put all sort of living creatures until he decided, in a orderly manner, to create an almost perfect living being,…man.

I used the word "almost" in purpose, to indicate the fact that nothing is perfect as God is. He made the whole universe almost perfect to give man a chance to improve it as his own knowledge grows in his continuous search for answers about the existence of everything around him. The latest advancement in sciences also favors this view as it was observed the existence of asymmetry ranging from subatomic particles proved by quantum mechanics to configuration of galaxies as proved by the latest cosmological studies.

The bible also show us that as a creator God is also the greatest artist and like any other artist, while making something from nothing, many times he changed his mind. Natural examples are the extinction of dinosaurs and some vegetal species and biblical examples are when Moses wife asked God to change his mind about killing her husband and other we can read in Jonah's book, when God, expected to be convinced by men against destroying Nineveh.

Nowadays, using our well advanced technology it was found that in thousands, or even millions years to come, our world is going to an irreversible end probably by fire, like the bible said.

Many scientists think that the whole universe will end in a singularity described as big crunch or like a big explosion (of our entire galaxy), in a similar way than the moment when God decided to create it billion years ago.

The most popular theory about the creation process has been described by scientists and accepted by almost everybody as the Big Bang (A term created by the English scientist Fred Hoyle which ironically he did not like it much).

In this book we agree that there is only one single process of creation for the whole universe, when time and matter started the complete process of evolution to the formation of water and minerals and with these, of life and man, all by the hand of God.

About the time of ending, in the past we used to picture our Earth as the main thing in the universe surrounded by the sun and all other stars,. But we learned now that our planet is only an insignificant dot in the Milky Way Galaxy which in turn, is a tiny speck in the enormity of the universe bringing us to think that when we talk about creation of life on Earth we might me talking about a "local" situation. Then the biblical end of life in the universe could be equal to the physically calculated for our Earth, affecting only our planetary system, and probably extending to our entire Milky Way Galaxy and even also the neighboring ones, but this might not have any influence in the existence of the whole universe…Unless God decide to do that…

Galaxies like our Milky Way, which includes our beloved Earth, will grow to maturity and will die to be regenerated into new matter at the edge of the universe…

We have learned about the big catastrophes in the world since its creation. A number of them, like the Ice Ages are not specifically mentioned in the bible but are indicated in the rock formations. Others might be correlated to some passages of Genesis on the bible and there are others like the flood, that are mentioned in the Book but no evidence was found on Earth away from the biblical land.

Astronomers tell us now that there are over 2,000 asteroids with sizes of about half a kilometer across, threatening to hit our Earth and there might be thousands more to come, smaller, but as dangerous as the 100-meter diameter asteroid that hit Tunguska in Siberia and flattened about 1,800 square kilometers of forest on the early nineteen century.

Some geologists believe that an event similar to this, but of larger proportions might have caused the extinction of the dinosaurs and there is a general belief that a big one grazed the Earth setting the origin of the Moon when the Earth was still in its first stages of formation.

If we start considering in detail our situation in the universe, with all those threats surrounding us, we will understand that our own existence is nothing else but a continuous miracle that can only be kept by the constant observation of a Creator.

Even considering all the natural defenses we were given, such as the gravity field to keep everything attached to the surface of Earth, or the atmosphere to filter and burn all incoming foreign bodies, and the later formation of the Moon, attached to our Earth to create tidal forces that generate clouds and the wind to move them, as well as many other forces that helped us to live safely through few millenniums, these associated facts makes it very difficult to believe that is the product of a series of connected accidents.

Considering all these factors and going back to the Book of Revelation 6.12, we could assume that what the astronomers described about the asteroids, might be related to what the Bible say…"6.12. *I watched as he* (The Lamb), *opened the sixth seal. There was a great Earthquake. The sun turned black like sackcloth made of goat hair, the whole Moon turned blood red and the stars in the sky fell to Earth, as late figs drop from a tree when shaken by a strong wind. The sky receded like a scroll, rolling up, and every mountain and island was removed from its place.*

How near are we from all these to happen?...No man has yet the knowledge to make a prediction on these events as we still have too much to learn about the mechanics of the universe. According to Hubble, as the universe is expanding since the moment of creation, our galaxy is moving away at increasing speed and even if the rate of expansion is decreasing, this means that there will be a moment that our galaxy speed will reach a critical limit to its destruction. We know now that our galaxy is moving at a speed of 600 kilometers per second and we know that the light travels at 300,000 kilometers per second. Somewhere between these, there is a critical speed when the mass of our entire galaxy might disintegrate in billions single unit particles which eventually could become energy to be incorporated into the universe's mass, but this will be discussed in more details on the following chapters.

Genesis starts with a careful but general description of God's creation showing that everything was done in progressive, well planned steps and in different stages called days.

We define a day as a period of twenty four hours when the planet rotates around its axle. This is a very intelligent way to measure time using God's given nature but in the search of perfection man has already found a more precise measurement using atomic clocks to define more precisely one second of the hour which should still be a part of an Earth's day measured in solar system units, but God's days were indeed different as they probably account for the existence of the whole universe.

We can read in 2 Peter 3.8. *But do not forget this one thing, dear friends: With the Lord a day is like a thousand years, and a thousand years is like a day…*

At the beginning the bible explains very clearly that water is the essence of our world. In the second day of creation God said, *Let there be an expanse between the waters to separate water from water…*

This cannot be more clear, water is the main component of our universe, water formed minerals which initially formed the core of Earth and surrounding it there are other mineral associations developed in the mantle where huge melting chambers known as magma, were and still are, the source of all igneous rocks that fill our landscapes.

These magmas contain a mixture of different molten elements with abundant water at very high pressure and temperature which periodically gush up to the surface of the planet forming new mountains and thickening our Earth's crust.

The residual waters from all these enormous chemical processes since the origin of our Earth also raised to surface filling the big existing basins to form the oceans.

When God talks about the creation of the Moon in the bible, explains that this happened after there was life in Earth, during the fourth day.

Most of the geologists might not agree with this idea, but we think this could well fit into the geological chronology and it could be that our Moon is part of the Earth torn off by a passing planet or asteroid from the area where the Pacific Ocean lies now. This event triggered the separation of land in different continents as we know it now, beginning with the continental drift and the formation of plates moving along the bottom of the seas.

Some geologists think that the Moon was a loose body captured by the Earth gravity, but most of us agree that it was a later addition to our planet system, another act of God working on the improvement of his beloved Earth. The question is when and how all these event happened…? This we will discuss later using the latest finds of our research on results from work done by well known organizations and scientists around the world.

Now we want to complete part of the teaching from the bible.

At the end of his evolutionary work, God said,

"1:26…Let *us make man in our image, in our likeness, and let them rule over the fish of the sea and the birds of the air, over the livestock, over all the Earth, and over all the creatures that move along the ground…*This was on the sixth day and then we have the whole cycle of creation complete. A perfect evolutive sequence from particles of light becoming matter to form our Earth with all its components, including man.

Going to chapter two, it starts…"Thus *the heavens and the Earth were completed in all their vast array.*

*By the seventh day God had finished the work he had been doing; so on the seventh day he rested from all his work. And God blessed the seventh day and made it holy, because on it he rested from all the work of creating that he had done…*"

God promised he will not put and end to man,…He formed the whole solar system as the home for his main creation then, till all the prophecies of the bible are completed, only man could do this to himself before the due time…But this is more for the theologians to be discerned…

Lets start from the real beginning…

# Chapter Two

## The Beginning

*GENESIS. The Beginning.*

1. *In the beginning God created the heavens and the Earth. Now the Earth was formless and empty, darkness was over the surface of the deep, and the Spirit of God was hovering over the waters.*

   *And God said, "Let there be light" and there was light. God saw that the light was good, and he separated the light from the darkness. God called the light day and the darkness he called night. And there was evening, and there was morning-the first day.*

These explanations of the bible also includes the latest Big Bang theory sustaining that universe was created from a hot spot some fifteen billion years ago, with concentrating matter that formed the proto-planets as it is explained in many different ways by a number of scientists.

Presently. there are new, very complex and elaborated cosmological theories which are mostly understood and discussed by a small number of experts who keep on feeding the rest of the world with their new findings and from there, almost everybody learn enough to talk about and make comments in their next social meeting, forgetting most of it shortly after to continue living their own lives minding their own business.

That is the way we live today, we know that our intelligence and knowledge has a limited range. We might become experts in a specific branch of medicine, physics, chemistry, geology, finance and many other sciences or business, but nobody can claim to have full knowledge of all sciences as it could be done hundred years ago.

Einstein was probably the most intelligent man on Earth during his life time, but he also had to admit being wrong sometime, like when he tried to explain that the universe's balance of gravity by applying a "cosmological constant" to prevent it from changing in size.

The Old Testament has reliable records that tell us about our world's creation and evolution into what is now our solar system and there might be more data hidden on those words that will eventually lead us to find more about the whole universe.

There were always some criticism done about the vagueness of some of the expressions found in this book but,…how could we explain all what we know now about our world to a primitive man in pre-Christian Israel?…The bible did it simple and in a very clear way.

It starts explaining

*…In the beginning God created the heavens and the Earth…"*
That indeed is the beginning of everything and is what we all wonder about. We might eventually end up finding how our Earth was created by the will of God. We know now he used all the best physical and chemical components to give us the world we live in but, the origin of heaven we still find hard to explain, because until today and even accepting few guesses from experts, nobody knows exactly what heaven is.

So now we will continue reading and analyzing

*…"Now the Earth was formless and empty…"*
All right, we all know here that God was talking about the whole universe or at least our solar system…But,…how can you explain all that to a primitive man that hardly knew the area where he walked about?…

Now we know a bit more about physics and the term formless and empty can be easily understood as a gaseous phase in the creation, it does not clarifies whether this gas was hot or cold but God expects us to learn more in our process of evolution into wise men.

Edwin Hubble found in the late 20's that all galaxies are gradually moving away from each other proving that the whole universe has been expanding continually since the moment of its creation. (This was an indication of Einstein's mistake about his cosmological constant).

To understand better what happened at the moment of creation if we reverse this process, we will probably end up assuming that, at a given time, the whole universe was so close and so dense that only God could have given it the spark of life and gradually let it grow from what the bible said to be…

*"formless and empty,"*

…That is the proto-universe that would gradually evolve from a cloud of gas, most probably helium which the most common component of the universe, through a concentration of matter into proto-stars with their planets and by expansion into the wonderful world we are allowed to live in.

The following words are really interesting and give us a few important clues to develop a scientific theory of creation based on the bible,

…*"darkness was over the surface of the deep…"*

This is a hard one to crack. It gives the feeling that the huge gas cloud had no boundaries but there is a magic word coming to our rescue, this is the word "water" as the Book says,

*"…darkness was over the surface of the deep, and the Spirit of God was hovering over the waters."*

So we can assume that at the beginning we had a gas cloud and creation started when this cloud cooled down (Or warmed up if some theorists prefers), to convert this gas into water, the main component of all matter in our Earth and the most necessary ingredient for the origin of life.

Hubble's theory about the expansion of the universe is now widely accepted with different modifications but also later on, Albert Einstein proved the expansion theory mathematically and more recently NASA scientists found enough proof to sustain all these convictions concluding that the Big Bang might have happened some 15 billion years ago.

After an immense explosion, fragments of matter concentrated about their own gravity centers forming ours and other solar systems. The matter was a mixture of gases heated to extreme high temperature and pressure taking shape and starting to rotate around a central more powerful mass,(the sun).

Our Earth was a small part of all that colossal formation process and started forming about ten billion years after the sun by different and very complicated physico-chemical conversions while it was cooling down and growing to get to its present size.

The best simple way to imagine this complex process is to remember that gas could become liquid under pressure and the liquid becomes solid when cold and further more water is the only substance that can be easily converted in any these three different stages as there are represented by the oceans, lakes and rivers, and as gas in moisture, steam and clouds, or as ice. We also now that helium is the most common gas existent in our Solar System as is the most important component of the sun itself and it is closely related to hydrogen which, with oxygen are the main components of water.

Then we can conclude without any doubt that the presence of water was the main component for creation…The bible said it…If we read, 2. Peter 3.5. *But they deliberately forget that long ago by God's word the heaven existed and the earth was formed out of water and with water.*

As we follow reading on the Book of Genesis we find:

*3. And God said…"Let there be light" and there was light. God saw the light was good, and he separated the light from the darkness. God called the light day and the darkness he called night. And there was evening, and there was morning-the first day.*

After the superheated proto-planet started increasing in volume, slowing the speed of rotation and then cooling down, water in form of steam started to condense and pour down to earth forming lakes and clouds which continued lowering temperatures and cleaning the atmosphere, while water from inside the earth, was liberated from its different magmatic chambers existing inside the mantle bringing more minerals in solution and started the formation of a crust. These colossal sequential processes led to the formation of the first islands which are identified now as the roots of all continents and are known as the greenstone schist belts.

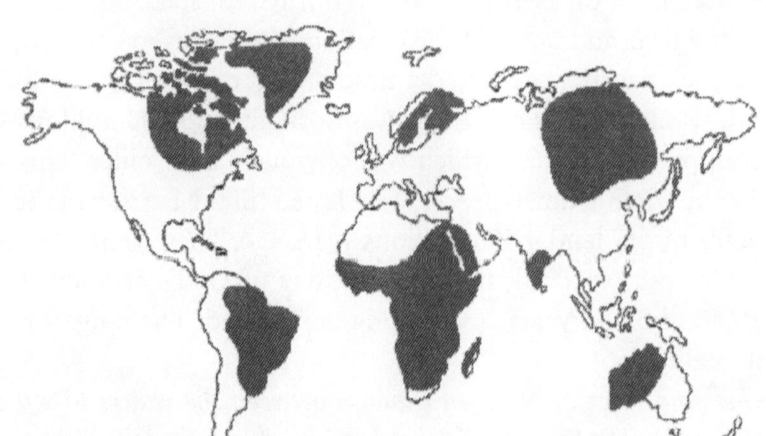

Fig.1 The areas shaded in black show the greeenschist belts of Archean age. The rooth of the continents

The oceans surrounding these islands were the residual effluents of all these processes. Huge amounts of magmatic material and liquid waste were released through fissures to start forming not only the oceans but also the primordial atmosphere composed of carbon dioxide, nitrogen, water vapor and other gases.

With the help of rains on repetitive cycles of evaporation, cloud formation and again rain, the surface become cold enough to allow life to grow.

All this was continuously interrupted by hurricane type storms, black dust storms, earthquakes and volcanic eruptions indicated today in many rock formations.

Then finally, this chain of volcanic eruptions accompanied by earthquakes and followed with dust clouds that covered all the planet, was later cleaned down by the repeated rains that cooled down the surface of our planet and the result was an original Earth.

Now we must go back to the bible descriptions and must consider the meaning of the words light and darkness. God called the light day and the darkness night on the first day. We know now that the sun was already there from the beginning as it is our parent star.

From the time and point of the creation of the universe to our present days, the scientists calculated a time period of about fifteen billion years. We also know that our Earth was born about 5,000 million years ago, being part of our sun which was obviously a bit older. This, followed by the cooling down process that lasted till the Earth had the first appearance of dry land 4,600 millions years ago. These general outline of events take us to think that the first day of God's creation, lasted some 10,400 million years, a very long day indeed. The longest of the biblical week.

*...God said, "Let there be an expanse between the waters to separate water from water" So God made the expanse and separated the water under the expanse from the water above it. And it was so. God called the expanse sky. And there was evening, and there was morning—the second day.*

Here again the bible explain very clearly that the first part of God's creation was the creation of water. That was the primary component and it existed in at least two different stages during the period of the planet formation.

The *expanse* (After named sky) was used to separate "water from water". then we can think that the *expanse* is the atmosphere as God created it, separating and protecting the surface of Earth.

Geological investigations show that the whole Earth was covered by ice in the late Precambrian. Could the bible be referring also to that?…That could be a very easy way to explain a complicated process that many of us still do not understand on the present days.

Now, following Hubble's theory of universal expansion, the Earth, like the rest of the universe started moving away from the sun and all other surrounding planets while it was also growing in volume like the points marked in a balloon when being inflated.

This expansion effect, forced all planets to rotate slower (as there were growing bigger) and go around the sun on a wider orbit, starting having nights and days as a process of natural evolution into a more stable planet with the development of the Earth crust. Rains helped to settle the dust down and increased the volume of rivers to transport sediments with the subsequent evolution of sedimentary stratigraphy.

*…And God said, "Let the water under the sky be gathered to one place, and let dry ground appear". And it was so. God called the dry ground "land" and the gathered waters he called "seas". And God saw it was good.*

The Precambrian era starts about 4600 million years to 570 million years ago, a span of over four billion years in which our planet's crust developed giving origin to the continents. This is the time when cooling of the surface became bearable for the creation of primitive life. This second day we could try to correlate with the Precambrian Era that lasted 4,030 million years, still a very long day but very much shorter than the first. We are now talking about reaching some 570 million years ago

> *...11. Then God said, "Let the land produce vegetation: seed-bearing plants and trees on the land that bear fruit with seed in it, according to their various kinds". And it was so. The land produced vegetation: plants bearing seeds according to their kinds and trees bearing fruit with seed in it according to their kinds. And God saw it was good. And there was evening and there was morning—the third day.*

In the geological sequence there is not a clear transition between the younger rocks of the Precambrian era and the older of the Paleozoic era also known as Cambrian period.

This sharp step shows there is a big, non-recorded event, that covered most of our planet before the first life forms started to develop in the Cambrian period and this could be related to another ice age. So this third day mentioned by the bible fits very well with the beginning of the Paleozoic era when all sorts of life bloomed over the Earth.

Our geological time scale describes that at in the Cambrian period, which is the older sub-division of the Paleozoic, marine animals with mineralized shells appear. There were trilobites, echinoderms, brachiopods, mollusks, primitive graptolites, corals and some primitive fish.

Later, during the upper periods of the Paleozoic such as the Ordovician, Silurian, Devonian, Carboniferous and Permian, other living creatures were formed in the sea and plants appeared in land.

Perhaps we could tie some biblical passage to the Carboniferous-Permian period after the formation of the well known coal seams around the world. We know now for certain that coal come from buried vegetation. It all happened at about the same time in all the continents meaning that some main event affected the new forms of life at that time.

(Psalm104:30)…*When you send your spirit, they (The creatures on Earth) are created, and you renew the face of the earth…*

This could bring us to think that God might have changed his mind about some type of vegetation or he rather had it all planned from the beginning and decided to provide man with some natural fuel giving him time to learn to produce his own. Here again there is a case were our Lord applied what we learned to call evolution. He converted some trees into natural solid fuels by a process of natural changes.

Obviously God liked his own idea and knowing than the intelligence of men would also grow into inventing the fuel engine that would bring the need of natural oil resources, then later on, in the Cretaceous, he did the same with other primitive forms of life to give man also a liquid fuel.

With only God's precision the necessary environment was created for the formation of oil around the world, using what he considered to be ill developed living organisms, he converted them into oil by creating an anaerobic environment that gave origin to the black shales after some natural process of heating under pressure forming the oil and gas as we know it.

God the artist, the creator, liked all this and knowing that men would get to the extreme of killing each other for the possession of those oil deposits and perhaps, willing to teach man to share the good things, he repeated the same process few other times in different parts of the world mostly on the Cretaceous.

The majority of the economical coal deposits throughout the world were probably originated during the third day which could be correlated with the beginning of the Carboniferous to the Permian some 250 million years ago. This third biblical day, starting in the Precambrian and finishing in the Permian, lasted about 320 million years.

> …14. *And God said, "Let there be lights in the expanse of the sky to separate the day from the night, and let them serve as signs to mark seasons and days and years,* 15. *and let them be light in the expanse*

*of the sky to give light on the Earth. And it was so. 16. God made two great lights.The grater light to govern the day and the lesser light to govern the night. He also made the stars.*

*17. God set them in the expanse of the sky to give light on the Earth. 18 to govern the day and the night, and to separate light from darkness. And God saw it was good.19 And there was evening, and there was morning—The fourth day…*

Here it seems that the writers of the bible took one step back in the process of creation as they had already said in Chapter 1:

*1.11 "Let the land produce vegetation: seed bearing plants and trees on the land that bear fruit with seed in it, according to their various kinds"…*

If in 1:11 we already had the land with plants and trees, then surely we also had the sea, rivers and rain. We also know that these trees need the sun to generate the cycle of life by means of photosynthesis, we also need good air and water, so…What is this all about the creation of the Sun when we know it was our parent star and it was there before…?

…Or at least at the same time that our Earth…

Then,…why explaining all that again about the Sun and the Moon in 1:14?…

We have no doubt that the bible is the perfect book, created by God, using some chosen people to write it down, so we cannot even think there is a mistake in the writing, we have to adjust the interpretation.

Then we must rather review our geological time scale.

If we assume that the fourth day starts at the beginning of the Triassic period of the Mesozoic we find that the mollusks were the dominant invertebrates in the sea and the land was populated by dinosaurs and plants such as ferns, cycads, ginkos, rushes and conifers but, suddenly, for some not very clear reason, all life was cut down in land. This event was not specifically mentioned in the bible but we know of it by all the fossils found in different parts of the world during this geological time.

Dinosaurs reached mass extinction throughout the world at the same time along the boundary between the Cretaceous and the Tertiary. Also invertebrates such as ammonites and ostracods were affected, like many fish and plants.

These well preserved fossils are a clear evidence that whatever was the reason for their sudden deaths, it happened very fast, as most living things on that time were almost buried alive. The proof of this is that some animal fossils were found having food in their mouths.

There were also many species of vegetation terminated during this period, but the chain of evolution was not cut there by their disappearance, instead, it gave place to the rise of the modern vegetation and animals later in the Tertiary.

The bible explains about the light of the sun to govern the day and the light of the Moon to govern the night. The existence of the sun was not mentioned before in the bible and St.Augustine explains in his wonderful book written many centuries ago, "The City of God"…"but the first three days of all were passed without sun, since it is reported to have been made on the fourth day…" and he continued explaining the meaning of evening and morning during these three days.

This could be understandably expected from a man of God that wrote such a monumental theology history more than fifteen hundred years ago, when people though that our Earth was the center of the universe, but we know now that the Earth is only a small part of our solar system and was certainly created after the sun.

If we read on the bible about day one, from verses 1:3 to 1:5 we read about light and darkness which are called day and night.

Also as on the third day there were trees and other vegetation in land, we have to accept now that the sun was already there and the only reason for it stopping shinning could be attributed to a world catastrophe and then, we can easily assume that the surface of the Earth was blackened by a huge cloud of dust or cinder...The end product of a major event such as a big collision with other planet and the origin of the Moon.

There are many theories trying to explain the formation of the Moon; one of them sustains that the Moon was an outer body that got trapped by the gravity of the Earth. Others argue that a comet might have a scratched the Earth in the very early stage of formation when all planets were still an undifferentiated mass. We support the theory that part of our Earth was torn away probably from where the Pacific Ocean lies now, leaving that great cavity on our planet and putting a large amount of dead massive rocks in orbit.

A third hypothesis talks about a near hit of our Earth with another body during a very early stage when the planets were still forming and the Moon is a left over fragment trapped by the Earth gravity.

Considering any of these possibilities, this catastrophic encounter would have affected badly the Earth rotation killing most living matter as dust and cinder might have covered the sunlight while being blown by powerful winds over the dying animals and trees. An event of such magnitude might have lasted at least, few thousand years.

These could be related to the reference of the bible about the revival of the sunlight and the beginning of the moonlight.

After everything quieted down, and as we got further in the Tertiary, mammals became abundant with the outcoming of rodents and primates.

This fourth day full of great events lasted 190 million years to the end of the Paleocene Epoch still in the Tertiary Period.

20. *And God said, "Let the water teem with living creatures, and let birds fly above the Earth across of the expanse of the sky."* 21. *So God*

*created the great creatures of the sea and every living and moving thing with which the water teems, according to their kinds, and every winged bird according to its kind. And God saw that is was good. 22. God blessed them and said,"Be fruitful and increase in number and fill the water in the seas and let the birds increase on the Earth." 23. And there was evening, and there was morning—the fifth day.*

Following our geochronological list after the Paleocene in the Triassic Period we have the Eocene, Oligocene and Miocene Epochs, all together lasting about 40 million years. In these Epochs the dominant species are the mammals with the appearance of the whale in the Eocene,(Notice that the bible says)

*…"21. So God created the great creatures of the sea…",*
until the end of the Miocene when all modern birds completed the present animal kingdom stock.

*…"and every winged bird according to its kind"…22. God blessed them and said, "Be fruitful and increase in number and fill the water in the seas and let the birds increase on the Earth…"*
Then, according to the bible the modern birds creation is the final touch of the fifth day and our geochronology agrees that modern birds were developed during the Miocene.
The other interesting point to rise here is when we read…

*"Let the water teem with living creatures, and let the birds fly above the Earth across the expanse of the sky…",*

This is a purely evolutionist concept, the water gave birth to living creatures and from the living creatures of the sea, birds were created. A proof of this could be observed on the birds legs, they still conserve remaining of scales in the skin while the duck, swam, goose and other *palmipeds* are the intermediate creatures that kept very close to the water.

Like God took a rib of Adam to make Eve He also took parts of fish to create birds and reptiles....Oh God...You are the greatest evolutionist...

24. *And God said, "Let the land produce living creatures according to their kinds: livestock, creatures that move along the ground, and wild animals, each according to their kind" And it was so.* 25. *God made the wild animals according to their kinds, the livestock according to their kinds, and all the creatures that move along the ground according to their kind. And God saw that it was good.*

26. *Then God said, "Let us make man in our image, in our likeness, and let them rule over the fish of the sea and the birds of the air, over the livestock, over all the Earth, and over all creatures that move along the ground."*

27. *So God created man in is own image, in the image of God he created him; male and female he created them.*

28. *God blessed them and said to them "Be fruitful and increase in number; fill the Earth and subdue it. Rule over the fish of the sea and the birds on the air and over every living creature that moves on the ground."*

*Then God said,"I give you every seed-bearing plant on the face of the whole Earth and every tree that has fruit with seed in it. There will be yours for food.* 30. *And to all the beasts of the Earth and all creatures that move on the ground—everything that has the breath of life in it—I give every green plant for food. And it was so.*

31. *And God saw all that he had made, and it was very good. And there was evening, and there was morning—the six day*

Our geochronology could locate this day from the end of the Miocene covering the Pliocene to the end of the Tertiary and into the

Pleistocene Epoch in the Quaternary Period on and estimated period of 18.2 million years.

The records say that during the Pliocene mammals evolved to modern form and there are the first indications of australopithecine species, the forerunners of humanity.

During the Pleistocene, already in the Quaternary and only 1.8 million years ago Homo sapiens evolves to spark the starting of human history.

So far there is no story of Adam and Eve, no paradise, no special rules. Man was created at the end of the line of all animals to live in this world, so carefully planned, improved and completed to the minor detail. God liked man best and allowed him to go freely around the Earth to live and reproduce like any other animal on this planet, the only difference was that man was going to have the responsibility to rule the world…

*"Be fruitful and increase in number; fill the Earth and subdue it…"*

God orders were to be fruitful and fill the Earth with descendants and men did as were told.

So far the bible talks about perfect evolution, animal life start from very primitive beings to the most developed species on Earth. We have made great progress since then but, we still do not know how far man have to grow to perfection before getting God's approval. It seems we are still far away from that target.

On the present days we have already gathered enough evidence to prove that mankind has evolved from a very primitive animal-like being, living in caves and fighting with the neighbors to protect their kin. Later, as he got more clever he started making alliances with others of the same strength to fight the more powerful until reaching today's way of living in well kept houses, in towns, using specialized people to protect their homes and forming part of a nation with people working to keep the wars away from home.

The bible closes the chapter one at the end of the six day, when the world is already mounted and functioning as we know it.

Now let's suppose that we do not believe in God or in all what the bible says. Let's suppose that we are pure scientists and we learn that some fifteen billion years ago from nothing there was a huge spark that started the whole engine of the universe that contains our solar system, then we can try to explain how, but there will be no way to say why.

We can also explain that the Big Bang was an event too perfect to be generated by accident, had this big explosion been slightly less powerful, the expansion of the universe would have been less effective and in a few million years the universe would have faded away causing the whole system to collapsed back to itself. On the other hand, if the explosion had been slightly more violent the universe might have got disperse randomly away.

The latest work in astronomy established the existence of many galaxies like our own Milky Way, many with strong possibilities of having other solar systems like ours and then, life outside our planet is something not that impossible to think about. But there is no reason for the bible to talk abut that. The logical thinking for a believer is that if there are other populated planets the people there might have their own book about the same God.

We know now that life is the final product of a long series of complicated physical and chemical transformations that could only occur under very special conditions and during an exact amount of time and that cannot be attributed to meaningless accident then, even as unbelieving scientists we should need to discover God.

# Chapter Three

## Those Were The Days...

*2. 1 "Thus the heavens and the Earth were completed in all their vast array.*

*2. By the seventh day God had finished the work he had been doing; so on the seventh day he rested from all his work.*

*3. And God blessed the seventh day and made it holy, because on it he rested (or ceased) from all the work of creating that he had done.*

This is the last day of creation that could be correlated with the beginning of the Quaternary, starting on the Pleistocene about 1.8 million years ago with the ice ages and the evolution of the Homo sapiens, to the Holocene with the starting of the human history about 3,500 years BC. Total length of the seventh day to date is a bit more than 1.8 million years.

At this point we are prepared to accept corrections from cosmologists, astronomers, mathematicians and stratigraphists to refine these figures but the general idea of the creation week could be summarized as follows:

| DAY | YEARS x 1,000,000 | MAIN EVENTS |
| --- | --- | --- |
| First | 15,000 | Creation sparks. God creates space, time, light, matter and instants after, evolution. Our Solar system is born. The Earth starts orbiting around the sun. |
| Second | 4,600 | God created day and night. The Earth is primarily formed |
| The first Ice Age | | (Precambrian Time). |
| Third | 570 | God created the land and the seas and provided vegetation. The Earth's crust started forming.(Paleozoic Era). |
| Fourth | 250 | God re-created the light of the sun and the Moon after the emplacement of the Moon around the Earth's orbit. This caused a main catastrophe in Earth. Many forms of life become extinct to give room to others. (Mesozoic-Early Tertiary). |
| Fifth | 60 | God created birds and the great creatures of the sea. The mammals became dominant with the appearance of the whale. Birds populate the Earth. (Middle Tertiary). |
| Sixth | 18.2 | The Earth is finally populated by all kind of mammals and at the end of this day God created man. There are australopithecine forms that evolved into humans. |

| DAY | YEARS x 1,000,000 | MAIN EVENTS |
| --- | --- | --- |
|  |  | (Upper Tertiary-Lower Quaternary). |
| Seventh | 1.8 | God stops his creation and decided to rest. Adam and Eve are have been created at not specified time. Homo sapiens develops after the ice ages. Beginning of human history.(Quaternary). |

By measuring the length of each biblical day it becomes very difficult to find an explanation to their difference in length.

It is logical to assume that the first day was the longest. We do not know exactly when the Earth was detached from our parent sun to start all the process of mineral formation and differentiation into core, mantle, and oceans after cooling down many thousand degrees while colossal forces were acting to form the primal Earth, but it is safe to assume it lasted 10,400 million years.

During the second day our Earth was placed in its orbit and kept on cooling down getting the atmosphere and the formation of some islands with few freezing periods or ice ages that led to the creation of the first living forms detected about 3.5 million years ago. This day lasted 4,030 million years.

The third day was much shorter, with the formation of the continents, when the first trees started growing in land and it lasted 320 million years.

During the fourth day there were some major changes on the Earth with the creation of the Moon and the adjusting of the new binary planet system on its orbit around the sun. It was the final stage in the formation of the continents. It was 190 million years long.

On fifth day our Earth was pretty much as we know it now, it was 40 million years when most animal species re-populated the Earth.

The sixth was 18.2 million years. The climate was readjusted few times through few ice ages till reaching the optimum before the creation of man.

Now, we are still on the seventh day and by projecting these figures, we could end assuming that this day should extend to about three million years living us over one million to think about what is going to happen at the end.

One important detail we noticed while reading St.Augustine's The City of God, he correctly mentioned there that God did not talk about the evening of the seventh day. Back to the bible we found the same sentence at the end of the first six days…:"*And there was evening and there was morning…*", but on the seventh day this sentence have not been added.

All these lead us to believe that we are at about noon of the seventh day and unless God looses his patience with us, we still have half a day to go before the end.

If we compare these day length and we refer them to our geological time scale we obtain the following:

| Biblical day | Length years | Geological time |
| --- | --- | --- |
| Day 7(still going) | 1.8 million years | Cenozoic era-Quaternary |
| Day 6 | 18 million years | Cenozoic era-Upper Tertiary |
| Day 5 | 40 million years | Cenozoic era-Middle Tertiary |
| Day 4 | 190 million years | Mesozoic Era to Lower Tertiary. |
| Day 3 | 320 million years | Paleozoic era |
| Day 2 | 4,030 million years | Precambrian Time |
| Day 1 | 10,400 million years | Earth formation |

It may be some meaning for the length of each day but, as we know that these estimated figures are still to be refined by studying in more detail our "geological clock" we will not continue on this subject now but, as it will be explained in chapter seven, it takes 220 million years for the sun to have a complete orbit around the Milky Way Galaxy then if we look the length of our Day 4 on the table above and allow few more years due to the expansion of the universe and therefore the widening of the orbit…Could there be a connection in the length of the particular day…?…Then,…what about the other days…?…Well we need a mathematician to play around with those numbers.

On the description of the creation on the sixth day, the bible explains…

27. *So God created man in is own image, in the imagine of God he created him; male and female he created them.*
28. *God blessed them and said to them," be fruitful and increase in number; fill the Earth and subdue it…*

In this part the bible there is not talk about any special man made of clay and converted into a living being by the act of God. Further more, in 1:27 the bible explain very clear that man is the natural end product of all creation.

God had planned to create man all the way from the beginning and slowly and almost perfectly was shaping all the environment till it was right for men to live in comfort.

He also knew that man would be able to start using his brain to improve his way of living all the time, setting himself goals for creating his own small "paradise" surrounded by kin and friends.

The bible said that God enjoyed his creation all the way and to make it more complete he wanted to have some special company close to him then, he probably decided to create Adam an Eve in the way we were told.

Later on God was repeatedly deceived by man, starting with Adam, continuing with Cain and later with the whole human race on the times of Noah.

Now we have to analyze something about Noah's times.

The story told by Shem is that because all sins committed around, God told Noah he was going to bring the flood over the earth as they knew it at that time.

If we look the map of the area described in the bible we can see that it is vastly covered by sands of an old sea and it is surrounded by waters of the Mediterranean sea to the West, the Black sea to the North, the Caspian sea to the Northeast, the Persian Gulf to the Southeast, the Indian Ocean to the South and the Red sea to the Southwest.

That was the world they knew at that time. The bible could not talk about the Himalaya, the Andes, the Rockies and other major geological structures around the world because nobody had heard of such places existence.

Like the bible does not tell us now about other worlds in other galaxies, there was no point in telling the old people about men living in other continents. That would have spoiled the whole evolution of men and our history, we would have not needed Marco Polo, Columbus, Cabot and many other discoverers and all the adventures that conquered and colonized the world building the civilization the way we got it now.

For the same reason we can also assume that The Flood did not extend further than the Middle East, the Arabian Peninsula and part of Europe and North of Africa, excluding the Ruwenzori Mountains which are the source of the Nile and are taller that Mount Ararat.

After all, the Old Testament is a chain of historical events exclusively written for the Jews living in that biblical region at that time. The importance of this region is that it also was the birth place of most religions on Earth.

# Chapter Four

## The Moon Of The Fourth Day

*GENESIS 1:*

> *...14 And God said, "Let there be light in the expanse of the sky to separate the day from the night, 15 and let them serve as signs in the expanse of the sky to give light on the Earth". And it was so. 16 God made two great lights—the greater light to govern the day and the lesser night to govern the night. He also made the stars.*
>
> *17 God set them in the expanse of the sky to give light on the Earth, 18 to govern the day and the night, and to separate light from darkness. And God saw that it was good. 19 And there was evening and there was morning.—the fourth day.*

Using the bible as a guide we can read at the beginning of Genesis that, at the moment of creation the universe was formless until God decided to create the solar system, and with the sun (light) and separate day and night. That is a clear proof that the bible meaning is that the sun was there from the first day. At least before the end of that day.

On the second day, God separated "water from water" to shape up all planets and the Earth surface ending with the formation of the atmosphere. This was the Precambrian times when the Earth has some alternative periods of cooling down that ended with some of the known Ice Ages, followed by warming up periods till reaching the optimal point step by step, as everything was taking shape in a very distinctly organized way.

We are still looking for more evidence to make an historical review of the Earth's climate but geologists have found proof of the existence of different periods where most of our planet was covered by ice. These Ice Ages were followed by sudden warming indicating once again that all was a well planned process of evolution within each geological period or day of creation.

On the third biblical day, God let dry ground to surface above the sea water level just like oceanographers describe the first island formation, and then we had a difference between land and seas with land fully developed holding vegetation with seeds and fruits on the land.

This third biblical day describes that life was already on Earth at that time. Our geological watch could well fit this third day in the Paleozoic starting with a worldwide discontinuity at the beginning, followed by the continuous development of a new environment to support better adapted forms of life.

This day was of a tremendous growth, starting in the Cambrian and going through the Ordovician, Silurian, Devonian until the Carboniferous when God probably sensed something going the wrong way or He just decide to make a change…

In our world, all artists are creators and whether they are writers, designers, musicians or choreographers, most of them will tell that in some stage of their work, they experienced that something they have created starts growing the wrong way or the wrong size and even when their creators know there could be easily eliminated with a brush stroke, pressing delete, or re-drawing the plans, occasionally some of those ill-created things manage to survive, even becoming dear and difficult to be detached of. It just seems they manage to keep stubbornly alive.

God made us in some way like him, then we could assume that in certain steps of his creation he might also had the same experience about things going the unwanted ways.

What the bible say about creation on the fourth day we are going to prove right as we can fit this events within our geological clock.

This is the first reference the bible makes about the Moon and then, the reader gets the impression that light was re-created in a more formal way, with the sun marking days and the Moon nights.

In order to explain this rather complicated process of Moon creation we must first analyze all the main scientifical theories and see how we can make them fit in the book that holds all the truth but first, let's analyze the whole situation by studying the following factors:

1. *The Earth and Moon density are different, the first is 5.5 and the second 3.3* This difference is mainly attributed to the lack of iron in the rocks of the Moon, a typical component of the Earth's mantle and core. The mantle has a density of 3.4 and the oceanic crust 3.0. (This is a point that indicates why the Moon have a poorly developed core helping to prove that it is just a lifeless piece of the upper mantle with a bit of crust from our planet).

2. *Moon diameter is 3,476 Km(2,155 miles) and the volume is 21,990,000 cubic Km, which represents 2% of Earth's.*(The diameter of the Moon could fit inside the area presently occupied by the Pacific basin).

3. *Moon rocks contain fewer volatile substances, like water. Also is very deficient in compounds containing carbon, potassium, sodium and chlorine.* (If we assume that the Moon is a dead chunk of rock extracted from the Earth surface and upper mantle all volatiles might have evaporated or spread into space on the collision. This

type of volatiles are essential for different ore formation in Earth and could help to explain the lack of them in the Moon).

The lack of other compounds it is explained by geologists of PERMANENT (Projects to Employ Resources of the Moon and Asteroids Near Earth in the New Term) indicating the Moon was originated later, after the sun and probably the Earth were full developed.(This could fit with out theory that the Moon was extricated from Earth not earlier than the Precambrian)

4. *Earth axis is tipped 23.5 degrees from the sun axis. Earth equator is tipped 5 degrees from the original* and there is a 5 degree tip in the Earth related to the Moon. (This is a very strong observation in favor of the Earth collision with other celestial body and this also shows that Earth and Moon were not created together as Earth had a different equator than the original and the axis related to the sun has also been modified. There is also a displacement of the gravity center from the original center of Earth to the center of Earth-Moon system and it is named barycenter situated close, but out of the Earth).

5. *Earth and Moon have the same oxygen isotopes($O_{16}$, $O_{17}$, $O_{18}$), composition which is different to all other planets and meteorites.* This point help us to discard the possibility that the Moon was an alien asteroid captured by the Earth gravity on its way to the sun.

6. The *core formation of the Moon was estimated 4.1 billion years ago. Surface material 3.5-4 billion years ago.* Here things get a bit complicated. The core formation fits well if we consider that Earth started to grow form about 4.6 billion years ago so the formation of the upper mantle and lower crust was completed later, but the surface material should have indicators of a big event

that happened about 0.1 billion years ago.(There is one interesting sample taken at Tycho crater's rim that showed about this age)

7. *The rocks are mostly basalts and anorthosite of igneous origin.* The basalts are a characteristic of the rocks in the Pacific basin. The anorthosite is also an igneous rock existent from early Precambrian times on our Earth. (This could be another proof of the "paternity" for Earth. Other important point is that the minerals that form anorthosite are rich in water and crystallize at high temperature and pressure. If the Moon was formed independently it should be rather all basaltic.)

8. *There are no magnetic fields and no tectonics activity.* (This means that the Moon is a dead piece of rock with no core, as it possibly was an old craton before the time of departure from Earth).

9. The *Moon's composition resembles that of the Earth's mantle.*

   *Another important point is that geochemists of the University of Michigan found that the tungsten isotopic composition of the Moon, is consistent with hypothesis that the Moon was derived from the Earth itself, or from a colliding object with similar chemical composition.*

   *John A. Wood explains that both, the Earth and Moon mantle lack siderophile trace elements as presumably they were absorbed in drops of molten metal that sank to the core during the primary differentiation of the Earth.*(Meaning that the Moon and Earth mantle had a common origin but as there was not a differentiation process in the Moon, its fission from Earth must have happened on a later stage).

   A.E. Ringwood and coworkers point out that abundance of the siderophile elements Co, Ni, W, P, S, Se and Te are practically the same in the terrestrial and lunar mantles, meaning that the same process could not have developed independently on Earth and its Moon as they could never have the same volume, amounts of pressure and temperature inside large fractionating bodies.

The scientist Drake(1983), pointed out that there are different concentration of the siderophile elements P, Mo, Ge, and Re in Earth and Moon. These elements are more depleted in the Moon, meaning that as the Moon separated from the Earth's mantle it had to undergo a process of core formation which did not happen on Earth.

10. *Simulations studied by the geochemists of the University of Michigan established that a giant impact might have originated "phenomenally high temperatures of more than 10,000 degrees K.* This heat might be responsible for the formation of the magma ocean in the Moon and also could be the reason why some indicators have been erased, but even so, this heat only affected the surface of the Moon and could not help to the formation of a molten core. (The lack of water and confined pressure might have prevented these rocks from having any type of metamorphic process).

11. *The Moon is getting further and further away from Earth with each orbit. This is happening at a very slow rate.* This is probably a combined result of the present expansion of the universe and is not related to its formation.

12. *The Earth is 81 times as massive as the Moon and also the Moon was partially re-molten when it was formed, but it cooled off quicker undergoing lesser metamorphism than Earth. The lack of atmosphere brings no change due to weathering on the Moon and then there is no significant sedimentary depositions to create orogenic forces like we have on Earth.* We cannot expect the smaller body to

react of the same way in the process of evolution of a planet system. The re-melting could have not be of importance as other scientists found there are "crater scars" from asteroids dating 3.5 billion years. (Again this data could only be an indicator of the original age of the rock and not of the impact date as the lack of water and an atmosphere might not allow transformation of minerals the way we know on Earth).

13. *The oldest Earth rocks are 3.5 billion years old and are located in a very small area of the planet and most of the old Precambrian rocks are less than 0.5 billion years old, while in the Moon most of the rocks are over 3 billion years old.*

(Here again we have to apply the word evolution to describe the Earth orogeny that led to the formation of mountains and associated mineral deposits which do not exist in the Moon).

14. *Dr. Norman MacLeod from the Department of Paleontology of the Natural History Museum of London explains that paleontologists think that mass extinction occurred between the Cretaceous and Tertiary caused by asteroid impact on Earth some 65 million years ago.* This is the time(That could be extended further to about 100 million years), when dinosaurs and most of its contemporaneous fauna and flora went into extinction. A group of scientists have been studying in detail this Cretaceous-Tertiary (K-T) boundary in different localities of the world and by sampling sequences of strata over the K-T boundaries in Brazos River, Texas, Barranco del Gredero (Caravaca) Spain, and similar deposits at Israel, Tunisia and Mexico, they constructed K-T sedimentary patterns showing a complete record of biotic events during that time.

Walter Alvarez and a team of geologists from Berkley, California collected samples on the Cretaceous-Tertiary contact zone in a limestone in Gubbio, Italy. This limestone was formed in a deep sea layer 65 million years ago in what is now the Mediterranean sea. Luis Alvarez, a Nobel prize winner in nuclear physics and father of Walter analyzed some of these samples finding they contain 10,000 parts per billion of iridium, compared with a usual background of 10 parts per billion in other layers above and below the Cretaceous-Tertiary contact. There is no dinosaurs fossils in Italy by elsewhere dinosaurs fossils are found immediately bellow this contact.

It has been found that meteorites and asteroids release considerable amounts of iridium during collision with the earth.

They conclusion is that *"at that time the Earth biota underwent a period of at least hundred thousand years of environmental changes prior to the impact event and the biotic effect of the impact itself appear to be largely confined to the tropical and subtropical regions with high latitude biotas which experience few K-T extinctions. The existence of the high latitude areas, isolated from environmental change, together with the substantial number of extinctions that occur in the uppermost Cretaceous period, indicate that the K-T mass extinction was caused by a variety of factors. Those probably involved climate changes resulting from widespread volcanism and sea-level changes in addition to those caused by an asteroid impact.*

Here, we could fit our theory and assume that this mayor event is related to the formation of the Moon. Times might not adjust exactly now, but reviewing our geological "clock", we might be able to connect the Moon catastrophic origin with the disappearance of most of the Earth life and even more with the beginning of the continental drift and the development of plate tectonics as it will be explained later). More studies on the Moon rock samples could help finding possibilities of this occurrence.

15. *The Moon has larger "crater scars" from asteroid impacts in one side.* This means that larger bodies hit the Moon and certainly our Earth in more than one opportunity. Specialist that studied the Moon craters estimate that the time of these occurrences are more than 3.5 billion years ago. These events cannot be corroborated on Earth due to the constant dynamics of our living planet. This could indicate that the Moon was an independent body orbiting Earth at that time and in that case we will have to admit that the Moon extrication occurred in the Early Precambrian.

(On the other hand, if we consider that the collision occurred on or very close to the south pole, part of that surface could have been kept protected by the ice for million years until the anorthosites were exposed again after the big collision and the ice melting).

16. *In a presentation at the Lunar and Planetary Sciences conference in Houston, project scientists said that lunar gravity measurements by the Lunar Prospector spacecraft confirmed the existence of a small iron core 225 to 450 kilometers(140 to 280 miles) in radius.* This latest find still allow us to think that this iron core could be a product of concentration of existing banded iron formations, formed on the sea near the continents, during Precambrian times.

The results of these scientists also confirm that the impact occurred after the Earth 's iron core was formed, ejecting rocky, iron poor material from the outer shell into orbit.

Now with all this facts from science we could move on in to develop some theories on the origin of the Moon.

# Chapter Five

## Origin Of The Moon

The main theories on the origin of the Moon are:

A). Capture Hypothesis.

This is an old hypothesis that suggested that the Moon was fully formed elsewhere in the Solar System and was subsequently captured into Earth orbit.

This theory lost credibility after the first rock samples of the Moon showed there is a close kinship in the Earth-Moon system. Also it was difficult to explain the capture process considering present orbit, lost of velocity and dynamics of gravitational circumstances.

B). Fission Hypothesis.

This hypothesis was presented by Charles Darwin, about 100 years ago. It explained that the Earth was spinning so fast that it formed a prominent bulge at the equator. This bulge was growing like a huge pimple until centrifugal forces threw out into orbit the piece of Earth that forms now our Moon.

This hypothesis seem close as the composition of the rocks is similar but later, it was proved that to have the necessary centrifugal force the Earth should have to rotate every 2.5 hours and in that case it should be some evidence about it. Also the Moon and Earth are not in the same orbital plane and the whole Earth-Moon system is also displaced from the orbital plane of the sun, meaning that our Earth has been moved from its original position. There are also other geochemical factors that turn this hypothesis unprobable.

C). The Precipitation Hypothesis.

According to this model the Earth began to accrete from existing debris in the same orbit and within a short time it grew in size until gravity forces caused the release of some of the impacting bodies. In this process the surface temperatures would reached about 2000 degrees Celsius vaporizing any incoming objects, Then the hydrogen and carbon monoxide with up to 20% in volatilized silicates would have formed a ring around the Earth.

From that ring the Moon could have been condensed and grew in size by accretion.

The dynamics of this process are not easy to explain to reach the present stage in the Earth Moon system.

D). Double Planet accretion Hypothesis.

On this model there are two planetesimals colliding to form our planet while some debris end up in orbit. This debris eventually concentrates in a sphere becoming the Moon.

This model fails to explain the 5 degrees difference in the inclination of the Moon compared with Earth. If they formed together in the explained way they should orbit in the same plane of the ecliptic. Also Earth and Moon should have the same density.

E). Giant Impact Hypothesis

This model have presently most of the scientifical support and due to that it has been thoroughly investigated using the most advanced computer simulation programs which are rather complicated to explain in detail but, basically, the idea is that there was an "impactor", which have been described as a body of the size of Mars, that grazed our Earth producing enough debris to create the Moon as we know it.

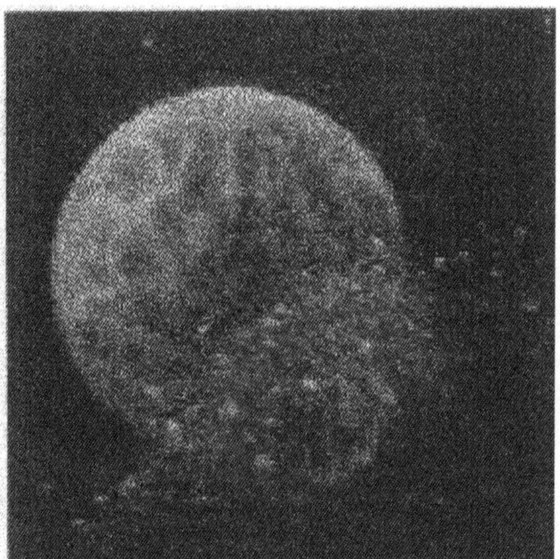

Fig.2 . This is a simulation of Earth being hit by a passing planet.

The mechanism is ruled by the gravitational potential known in the specialists circle as Keplerian trajectory and if the total energy (Keplerian plus kinetic) is positive then the debris will escape on a hyperbolic trajectory becoming a satellite, if not, then it will traverse in a closer elliptical orbit and fall back on the Earth's surface.

In order to get this material in the proper orbit there is two possible processes to consider;

1) The pressure gradients near impact site increase the angular momentum of material ejecting it into orbit or

2) the proto-Moon gains the angular momentum necessary to escape a re-entry by gravitational torque exerted by the bulge or discrete body on the planet.

Other factors considered for the assumptions made in this theory where, thermodynamics of impacts, viscosity of ejecta, dynamics of the process, numerical simulations, analysis of the chemistry of the impact and few other even more complicated methods to be considered any further in this work.

In all these experiments, they used temperatures of 2000K, considered for a very early Earth's surface temperature.

F}. Jupiter's influence Hypothesis.

In his book, Ice Age, Jon Erickson explains that Jupiter's gravity pulled an asteroid about the size of Mars out of the asteroid belt that lies between Mars and Jupiter. The asteroid grazed the Earth while the planet was still in a molten state. The asteroid gravitational attraction created a huge tidal bulge on the Earth surface and sucked a substantial amount of molten rock out of Earth.

G). The Mesozoic—Tertiary Catastrophe Hypothesis.

This is our theory, which is almost totally in agreement with the process explained in the Giant Impact Hypothesis but timely placed in a most recent age. It is also in agreement with an all theory sustained by the American astronomer W.H. Pickering, who believed that the Moon is the missing part in our Pacific Ocean and the more important point is that it is also in agreement with our bible study.

There are two main proposals to locate this major event within the time of formation of our Earth and both are responsible for many living species mass extinction giving way to the growing of more developed organisms, while continental drift was taken place leading to the formation of the oceans crust.

The first alternative is that this catastrophic event that ended with our Moon in orbit, might have happened during the Mesozoic (Jurassic-Cretaceous)-Tertiary time giving way to the immediate development of the continents.

The second possibility, reached after considering the age of the samples taken from the Moon by the Apollo missions, could be that the collision might have happened about 3.8 billion years ago, in the Archean, during the early Precambrian giving way to the development of an Ice Age that froze the whole planet.

The difficulty in accepting this second possibility is that, if this was the case, considering the writing of the bible, there was life developed much earlier on Earth.

Now we know that the Moon is a "dead piece of rock" meaning that unlikely our Earth,(which is probably the most vainglorious planet of the solar system as it keeps on putting "make-ups" by orogeny, sedimentation and metamorphism, over all scars caused by tectonism or asteroid impacts), in the Moon there is not a clear difference between core, mantle and crust like we have in our planet and there is no traces of any major sysmic or tectonic activity which are the common expressions of a living planet.

If we support the idea that if the Moon was "extracted or knocked off" from Earth when it was still in a molten, plastic stage, it should be expected to behave like its parent Earth with the iron dropping down to the center and the radioactive minerals generating heat to develop at least, a well defined core-mantle differentiation. But. this indeed, does not fit with the present Moon composition.

The fact that the Moon is a body composed of two main, very different types of rocks namely basalts and anorthosites, (These rocks were named respectively "mare or maria" and "terra or highlands" for the aspect they gave of "sea and land" on the first photographs).

Following the descriptions done by the experts we conclude that there is no "life" on those rocks as they have suffered little or not metamorphic changes throughout all events that happened billion years back.

This bring us to think that the Moon is the piece of our mantle with a part of crust attached that we are missing in the Pacific Ocean. This piece was probably located on, or very near our South Pole.

As the theory of the Giant Impact also talks about some after spinning of our Earth, it is to assume that this near collision was from an outer body coming in the same general direction, like a car being hit on the side by an overtaking faster car but,…what about if the collision occurred with the impactor coming in opposite way and hit, or grazed our Earth in an opposite direction to the rotation…?…Like a car coming from the opposite direction and hitting in the front mudguard…

Let's suppose that a foreign celestial body was coming in the opposite direction to the Earth rotation and our planet "mudguard" was located where the Pacific Ocean is now, but considering that at that time it was on or very close to the south pole. Then, the near collision produced the "dent" now filled with sea water and the piece of our Earth that formed the Moon would have flown into space and remain trapped by the Earth's gravity.

In that case we think possible that the impactor, might not have needed to be as big as Mars, but this we will leave to the experts in simulations to find out.

This bottom-side collision might rather slowed the Earth rotation and probably caused the swinging of the axial angle few degrees away and indeed had catastrophic effects over every existing thing on the surface of Earth.

The poles were moved away crushing ice and producing immense waves in the sea while in land, magmatic extrusions brought up by earthquakes, ejected millions of tons of lava, cinder and ash that blocked most of the living matter. That was the end for life on Earth for thousand years until the new Earth-Moon system found the new gravity center and managed to stabilize in the new position around the sun.

Now, based on the present results of the Moon mapping and sampling interpretations, we must consider the second possibility, we have to admit that if the evidence of Moon craters created by asteroid impacts are confirmed to be 3,5 billion old and older then, we will have to assume that this colossal event really happened in the early Precambrian. But still, we consider that to be a piece of the old Earth that was originally located at the south pole. With that, we assume that these rocks were preserved intact by a thick ice cover until the moment when the Moon was put into our Earth's orbit.

Obviously this theory cannot be applied to both sides of the Moon. The mare basalts are believed to be younger in age than the anorthosite rock that form the highlands in the far side. These anorthosites were the original surface rocks in the "Old Antarctic continent" that had already suffered the intensive action of meteorites like it happened in Mars, until the "big one" broke that enormous piece away from Earth.

The newly exposed face of the Moon could only show meteorite scars only after the collision but, if we consider that these meteorites were loose fragments of the same rocks that fell on the face of the newly formed moon, there might not be a clear record of the age of those impacts.

Going back to the origin of our Earth, we have to consider an initial faster speed of rotation generating angular forces that would probably concentrate the formation of land around the poles.

The Acosa gneiss found in the NW territory of Canada measured to be 4 billion years old suggesting the formation of the crusts started about that time on the far North…Could be that the location of our first North pole?…In that case the South Pole should have been in what is now the south Pacific Ocean.

In his book Marine Geology, J.Erickson describes an area in the North American continent known as the oldest (and probably biggest), on a region between Canada's Bear Lake and the Beaufort sea. There might lie the roots of an ancient mountain range running though the basement rock, formed by the collision of an old megacontinent named Laurentia (Consisting of the old North America, Europe and Asia continents pressed together) with an unknown land mass existent more than a billion years ago. This land mass was clearly located in what is now the Pacific Basin.

Marine geology also revealed that the older ocean crust in the Pacific is of Jurassic age and that at the end of the Cretaceous all water receded from land, as sea levels dropped and temperatures began to fall, ending with all tropical life and the existent marine dinosaurs such as ammonoids and rudists.

There is strong evidence showing that at that time the continents were welded together on the Atlantic Ocean side, with repeatedly openings and closings of small sea lakes, but with a single huge ocean on the Pacific Basin.

With all these evidence, how could we explain the existence of a very unbalanced Earth at the end of the Cretaceous where all continents were clustered together on the Atlantic side living a huge hole full of water in the other side?

Presently we have two thirds of the continental landmass in the northern hemisphere and below the equator there is only 10% of landmass on 90% ocean but, all was balanced when the Antarctic continent moved to the South Pole in opposition to the empty North Pole and Australia moved somewhere in between.

The ocean crust density is 3.0, compared to the mantle 3.4 and with the Moon 3.3, brings all the three to a very close association.

The Moon crust is mainly dusty with layers of outer broken rock called regolith. This crust and regolith are unevenly distributed on the entire Moon and there is a dominance of mare rocks on the nearside compared to the far side. Although the highlands occupy 83% of the Moon surface and the mare only 17 %.

This asymmetry on the Moon is another evidence to show that when it was placed in orbit it was already a piece of dead rock.

The little uranium and other radioactive elements left, mixed with the remaining of some banded iron formations contributed to the formation of the present, very poorly developed core.

Considering all these facts it is proposed here that sometime, between the end of the Jurassic and the end of the Cretaceous there was a big imbalance on Earth, created after the near collision of an asteroid of great size against our South Pole, this forced Earth to move its axis 23 degrees to compensate for the loss of mass while the continental mass on the Atlantic side was cracking and separating, moving with the mantle rocks towards the Pacific basin to fill up the huge hole, Antarctica moved South to form our new south pole and the whole process of continental drift started to take place with the development of the mid oceanic fissures infilled with molten mantle material that gave origin to the mid oceanic ridges and the formation of tectonic plates that we know now are part of our growing Earth.

In the mid Atlantic a ridge started forming filling an fracture that started opening like a zipper from south to north.

In order to leave such a big and dip scar as the Pacific Basin, all these series of events undoubtedly happened when our Earth was already fully developed few million years ago rather than on an younger Earth, when most of its material was still molten. If that was the case then, the plasticity of the heated matter would have smoothen all the features on Earth and we should not have the Pacific basin.

Moon sampling revealed so far that true age data for the basalts are between 3.8 and 3.1 billion years and for the highlands 4.5 million years. The interesting fact is that these impact marks indicate asteroid activities during a close geological time of about 700 million years. After that, there is no records of major impacts and most of the writers on this matter assume that very little else happened on the geological evolution of the Moon since 3.1 billion years ago.

Sampling on the impact sites proved that these "wounds" could be described as "self inflicted" as there are no mineralogical difference which could indicate rocks of foreign composition. It seems that those were pieces detached from the rough edges of the broken block in its process of rounding up in the shape of an sphere.

Finaly, in support of our theory the Moon was formed on Mesozoic-Tertiary days, there is a very interesting record from a sample taken from Tycho's crater rim in the mare site where one of the Apollos landed. It was dated 108 million years old making us to think that crater Tycho might be the product of a rock extracted from other Earth's rock formation at the time of the origin of the Moon. As meteorites disintegrate after impact it might not be possible to determine its composition but some remaining minerals could exist in the breccia that forms the regolith.

# Chapter Six

## The Pacific Womb

Now if we accept that the Moon is a part of our Earth broken away after our planet was already developed as the Bible said, then we can start checking things up by measuring the surface of our Pacific Ocean to see that the Moon would fit well in its basin. There is no possibility of measuring the depth as is expected that a cavity of that size would have been almost immediately filled by molten mantle material, continental debris and ocean water.

There are a series of major connected events that led to the evolution of our Earth the way we have it now, all of them perfectly planned by our Maker.

Their perfect synchronization make it impossible to think there are just a chain of luckily related accidents.

To make all these concepts easier to understand we must start explaining them on a series of progressive steps, following each link of the chain toward the origin of everything.

The first facts used in our theory were explained at the beginning of the last century by Alfred Wegener, a German meteorologist who proposed that the Earth was originally covered by an unique super-continent named Pangaea. This super-continent was later split horizontally in two huge super-continents, leaving the Laurasia super-continent in the Northern hemisphere and Gondwanaland super-continent in the south.

Laurasia contained what is now North America, Europe and Asia north of the Himalayas while Gondwanaland, contained South America, Africa, Madagascar, India, Arabia, Malaya, New Guinea, Australia and Antarctica.

From 360 million to 270 million years ago, Gondwana and Laurasia converged to form Pangaea, extending almost from pole to pole. This massive continent reached its peak size about 210 million years ago, during Mesozoic times. The primitive Tethys sea was suddenly overpowered by the huge Panthalassa sea, covering about two thirds of the planet.

The simultaneous breaking apart of all these super-continents was caused by Earth's internal forces that originated what is now known as the continental drift and the formation of new, large plates that move towards the continents on the bottom of the seas.

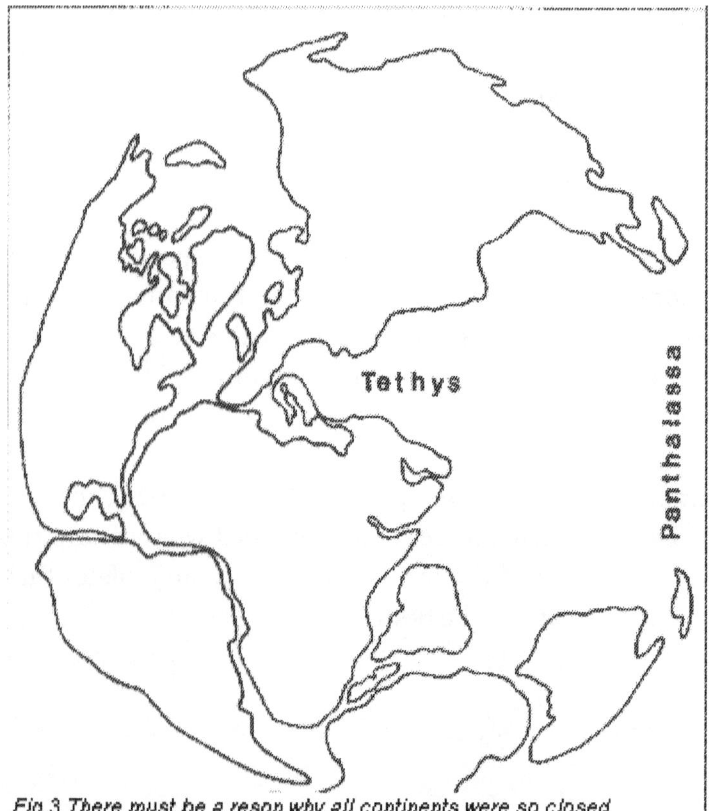



Fig.3 *There must be a reson why all continents were so closed toghether sometime before the Jurassic.*

This continental drift theory was ignored for many years until the concept of plate tectonics picked up the idea to explain the separation of the continents by rigid plates moving on the bottom of the seas from central fissures located in the mid Atlantic and Pacific Oceans, towards the continents where they sink into deep seated trenches that keep pushing this material to the mantle where it melts and regenerate as part of a new magma. These plates are cold and rigid, floating on the surface of a warm, less dense upper mantle, like chocolate bars floating on jelly.

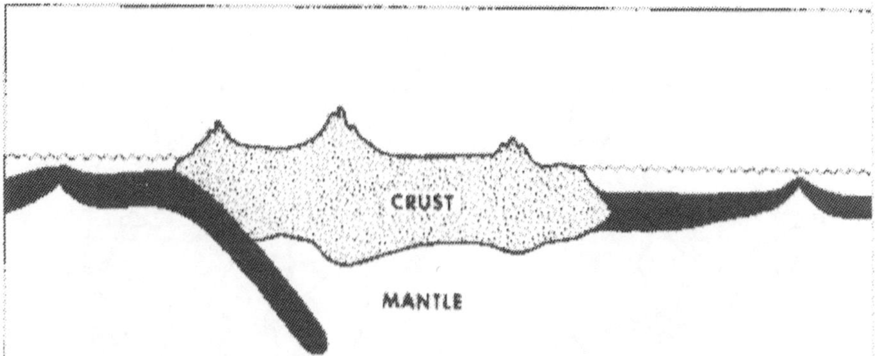


Fig.4. A West-East section of a continent and its relationship with plate tectonics. The plates at the left are in the Pacific and sink deep into the mantle to melt and re-generate.

Wegener discovered the very remarkable way in which all continents fit like pieces of a jigsaw on the Atlantic Ocean side. Their separation caused the development of a well defined central fissure that gave room to the formation of a ridge, which is still growing, splitting the sea in almost half.

On the Pacific Basin the picture is different as it seems there is no possibilities of any connection between the west of all the American continents with Asia or Australia. Remarkably there is something missing there as there are two main almost parallel ridges running in a northerly direction, one striking from Antarctica to the Western coast of northamerica and the other to the East of Philippines getting close to the coast of China. (These could be the fractures developed after impact and the fissures through most of the molten mantle material was poured into the basin to infill this enormous cavity)

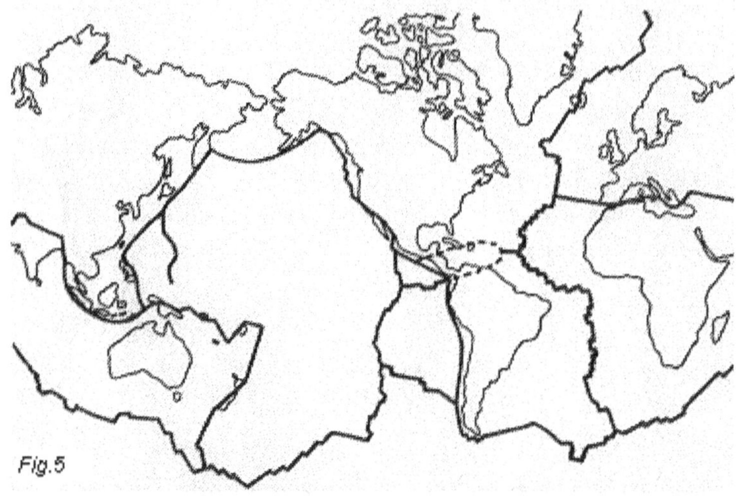


*This map shows the pattern of middle oceanic fractures that form the ridges where the tectonic plates start their movement towards the continents. Observe the difference between the single Middle Atlantic ridge compared with the almost circular of the Pacific.*

Wegener's idea gave origin to the generally accepted and well proved concept of Plate Tectonics, which explains with amazing precision, the self contained mechanism that keep our Earth alive.

The cycle starts close to the ocean ridge, with a thin ocean plate moving towards the continent. This plate gradually thickens by collecting material from the upper mantle underneath, and the accumulation of overlying sediment layers on the top, while moving away. The motion is created by heat convection currents originated in the mantle, pushing hot lava into the ridges and moving towards the edges of continental areas where dip ocean trenches are located. The plates are floating on this moving, heavier, viscous upper mantle material, growing in thickness until, eventually, the oceanic plate becomes too thick and heavy that cannot any longer remain on surface. It then bend downward and subducts beneath the continent into the Earth's interior adding more minerals to the mantle and providing more material to the molten magma for the formation of the new oceanic crust near the mid-oceanic ridges.

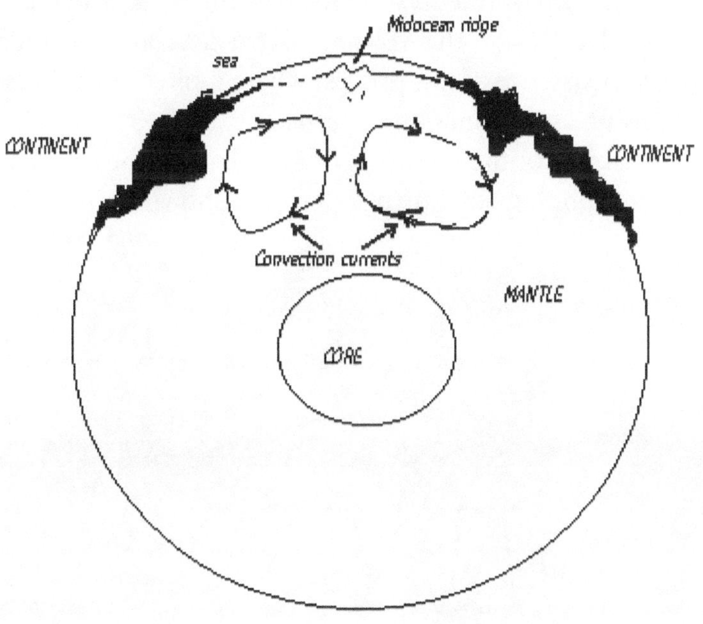


*Fig.6. The cycle of life of Earth. The convection currents move the molten minerals to surface where they form the different types of rocks that build the continents and move the oceanic plates from the midocean ridges, along the bottom of the sea. At the same time all sediments from the continents are washed down to the bottom of the sea forming a thick layer that exerts great pressure on the plates forcing them to subside under the continents as shown in figure 3. All this material is transported back to the Mantle where it re-melts and move along till it find the way to go again up to surface.*

This process is known to occur since the Precambrian on the north Atlantic plates but the oldest existing plates in the Pacific and Atlantic oceans are of Jurassic age. There are no records before that time and that could add some credits to the influence of the Moon for the Earth survival.

This is the way all continents were formed and is also the source of all known mineral deposits. the volcanic extrusions are the most clear example of the existence of this process where molten rock flows to surface along low pressure zones on the crust.

Another example of the evolution of our planet is the historical geology and geomorphology of North America as shown in figure seven.

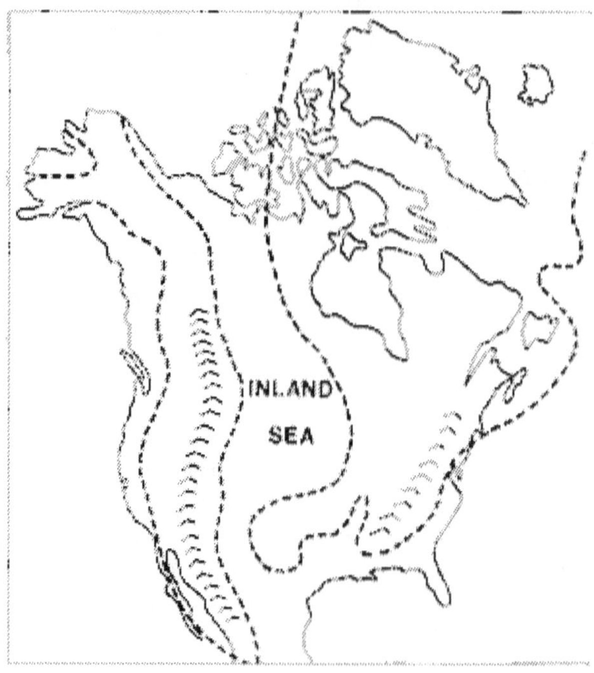


*Fig.7. A view of North America on Mesozoic days.*

During the Mesozoic era (Formed by the older Triassic, and younger Jurassic and Cretaceous periods), there was an intern sea in the west central portion of North America also covering what is now eastern Mexico, southern Texas and Louisiana. This sea also extended to South America, Africa and Australia and in Europe and Asia, it formed thick beds of dolomite and limestone that later formed the Alps and Himalayas.

At that time a western internal seaway divided the North America continent in two the Rocky Mountains and the Appalachian Mountains and to the end of the Cretaceous (and the Mesozoic), the water receded from the land dropping sea levels as temperatures began to fall causing the mass extinction of the warm climate life.

Marine geologists have found that during the Cretaceous period the oceans suffered a major oxygen shortage indicating there was major disturbance as the sediments at that time are characterized by very thick deposits of black shales.

These black shales, consist mainly of partially decomposed organic matter due to lack of oxygen. (The good news about these type of formations is that after million of years of compression and heating they became the largest oil fields in the world. Here again is God thinking about our future).

Later on, at the beginning of the Cenozoic era, in the Tertiary period, the high sea levels continued flooding all continents and it was noted there was a change in the deep ocean circulation with land gathering in one single area. These changes affected marine life and many species became extinct as the oceans covered 70% of the Earth surface to an average depth of two miles. The Pacific basin is the deepest, being at present one mile deeper than the Himalayas and its surface could have easily contain the Moon's diameter inside it.

Other factors to be considered in more detail are the siliceous ring in the sediments around Antarctica, extending close to the southern tips of the continents and the fact that the Atlantic sea floor consist predominantly of carbonates whereas the Pacific floor has mostly siliceous sediments with little carbonate present. As we know that the water composition is all the same in all interconnected oceans, this difference can only be attributed to the floor rock composition. The Atlantic basin being shallower might be covered by thick sedimentary layers while the Pacific basin being much deeper would expose the weathered rock. An interesting feature that could help to re-construct the structural changes in our Earth is the perfect semi-circular shape of the emplacement of the Aleutian Island between Alaska and Kamchatka. These islands are still growing along an active fissure that allows the flow of volcanic matter to the surface

Going back to consider the collision that originated the Moon, we know that such a tremendous impact generated high temperatures that vaporized water and some material but, if we consider that this block was part of an stable, anorthositic craton, forming a continent covered by a thick layer of ice, the resulting rocks put into orbit were rather unaltered with only some partial melting of the more affected impact areas that gave origin to some lava-like movements on and near the surface.

If we assume that one side of what is now the Moon was originally protected by a thick cover of ice at the moment of impact. The ice vaporized away mixed with other debris and vapors on collision exposing the fairly unchanged surface with the typical aspect of a "dead rock" as we know it now.

Other question we might ask ourselves is, with all these colossal processes occurring on the Moon we must expect that lots of heat might have generated on that small piece that got away from our planet, then, the whole Moon should be filled with metamorphic rocks like the ones we find in our Earth after some mayor geological movements.

But this is not the Moon's case because for any transformation in minerals to occur we require high temperatures in a confined place with high pressure and abundant water and none of those are or were available on the Moon so the only way that these rocks could have been heated up is as a result of falling meteorites which, depending on their sizes and angles of impact, melted and removilized the same old rocks, keeping the same or similar aspect that they had 3.5 to 4 billion years ago. Remarkably, these meteorites are of the same composition than those rocks found in the Moon showing that there were loose pieces of the "block" extricated from Earth that was "forced by nature" to become spherical by getting rid of all sharp edges.

The oldest ages determined by sample analyses on the anorthosites of the highlands are 4.5 billion years. All the impact formations are dated between 3.92 and 3.2 billion years. Later there are more impact records on the named Erastothenian-Copernican period was dated 1.1 billion and finally the youngest impact was found at Tycho crater, measured to be 0.11 billion years.

Specialist in Moon sampling admit about the difficulty they have to establish the absolute ages of the rocks as these absolute age records, are supported by tables of relative ages and those are not real statements of fact, they are rather statement of interpretation.

It was also found that from some samples taken on the Apollo 11 mission, the rocks thought to be older, were really younger rocks contaminated by an radiogenic isotope of lead that enriched the lunar soil.

The process followed to determine absolute ages from rocks is a very difficult task, consisting of measuring the minute amounts of radiogenic isotopes existing in the rock and then, calculate what amount of these isotopes has decayed into another element using a known standard rate. This technique assumes that the isotopic system of the rock has not been disturbed since it crystallized.

(The theory supported by the author here assumes that after this tremendous collision took place most of the mineral structures must have been affected some way and there must have been more than one phase of re-crystallization in some cases but due to the lack of components such as confined pressure and water which are essential for a metamorphic sequence there is no evidence about it).

In all these impact melts samples, age was determined using the argon technique, selected because the Moon samples have very fine grain.

The results showed that all have about the same age, between 3.2 and 3.9 billion years, meaning that geologically, the impacts occurred during a relatively short period of time.

The mare(or maria) of the Moon are made of floods of lava of basaltic composition similar to those on Earth and their ages range from 4.3 to 3.1 billion years old, compared with the Earth which is 4.5 billion years old.

Samples of the mare basalts show a complete absence in minerals that contain water. This could be an original condition or could be that water was lost after re-heating on a pseudo-metamorphic process. This re-heating phase on a local or even regional area could also resulted in the formation of volcanic craters. Other characteristic of these samples are that there are also depleted in all the volatile elements such as sodium, zinc, potassium and phosphorous but have a significant amount of titanium in form of ilmenite (Iron and titanium oxide), in opposition to the rocks in the Highlands which are rich in aluminum.

These basalts are indeed similar to those on Earth of the same age, with the only difference that all rocks on Earth had suffered a number of geological and petrologic changes through the ages.

Apollo 12 found younger lavas lower in titanium content confirming there are different ages of lava eruption…

…The question is…Are these lavas really crystallized billion years ago?…Could not be that the radioactive isotopes are kept unchanged because there were part of our Earth separated only a hundred million years ago and the lack of water, pressure and other Earth conditions after collision prevented these rocks from any major mineralogical changes…?

Any astronaut can tell you that to overcome gravity on Earth you need a rocket powerful enough to develop a speed of 11 kilometers per second or 25,000 miles per hour. This also helps to explain that the speed of the molecules of all gas forming our atmosphere are less than the terrestrial escape velocity and that is the reason why we are alive in our planet, but the Moon has 81 times less mass, therefore the escape velocity there is about five times smaller allowing to all gas molecules to escape easily into space. This might be another reason to explain the lack of changes shown on the moon's rocks.

The conclusion of some Moon experts is that there was a simple volcanic history with early lava eruptions highly rich in titanium and the later ones about 3.2 billion years old and since then all volcanic activity has suddenly died.

If this fact is finally proved right then our theory should be modified and applied to an earlier age, that is the big blow of a foreign body against our Earth happened in the Precambrian about 3.8 billion years ago.

If this was the case then, life in our planet was also developed earlier but we know that the atmosphere was still no right to support any form of life at that time.

On the other hand, this event could be the reason for a major drop on Earth's temperature that might have resulted in one of the main Ice Ages which have been proved to exist at that time but are still difficult to explain.

Using only the evidence we have so far from the Moon sampling, it seems there are better chances for this second possibility to have happened.

Then, according to our bible, there was life developed on land at that time as the third day of Genesis should fall inside the early Precambrian.

On the writer's opinion, without trying to ignore the results obtained from the rock samples of the Moon, the idea the Moon was originated on the Mesozoic seem to fit better and as all age determination done in different samples is still very inconclusive we will keep supporting it while trying to find more evidence to prove it.

Another examples are the samples taken of the ashes on Apollo 17 landing site. They were proved to be volcanic in origin and as the later event of a volcanic sequence there were supposed to be younger in age but resulted to be 3.5 billion years old.

Other find was that glass ejected from the mantle in the Moon, considered to de 400 Km deep appears to be largely unmodified from their chemical composition at their point of origin. This could be another point to direct us to think that the lack of many components like water and volatiles results in an unchanged rock that recrystalise on a similar way than the original, meaning there is no metamorphic processes and therefore no apparent age change.

These lava flows were proved to be unique for the moon, indicating a rapid discharge of material over a big area in a very short period of time giving this smooth, flat looking aspect to the mare. The total accumulated thickness of these lava is only few kilometers and many extensive areas do not reach 100 m.

Other Moon rock sampling revealed that the anorthosites that form the highlands of the Moon are the oldest rocks. (We have similar anorthosites in Loc St.Jean, in the Granville Province of Canada, also in Laramie, Wyoming and San Gabriel, in California which are part of the oldest rocks on Earth, over a billion years old).

The anorthosites in the Moon bear earlier crater marks caused by meteorite impacts. The resulting rock are called breccias, which are aggregates of a variety of highland rocks bound together by an igneous rock matrix of different grain size. Some of this impact melt rocks resemble mare basalt samples.

An interesting observation on these impact created rocks on the basalts and the anorthosites, is that their composition does not change, there are anorthositic with some basaltic resemblances or the other way around, in the basalts there are some anorthositic impacts meaning that these impacts are "self inflicted" by flying meteorites which were originally part of the Moon itself.

Other important point in favor of the Moon being a "stillborn" baby of Earth is that more than 70% of the Moon surface is covered by anorthosite. We know that for the formation of this rock the requirements are a confined, wet magma crystallizing under great pressure and high temperature. None of this conditions could have been found in any time on the Moon.

If we assume that the Moon was an existing chunk of cold and crystallized rock located where the Pacific basin is now, and that area was hit by a passing asteroid that put it into the Earth's orbit, we must expect there were some pieces of rocks of all sizes floating loose on the space, some falling into Earth but most of them going with the Moon.

Another consideration must be that this big broken piece of Earth had initially an irregular shape with edges that started braking loose as soon as the Moon found its orbit and began to round itself up like a small planet. If this block also contained other rock formations (Like those from the Precambrian detected in craters Eratosthenian and Copernican) their impact on the new satellite should reveal that age.

In his book "The once and future Moon", Paul Spudis explains that samples taken at crater Tycho indicate an age of 108-9 million years…, right on Cretaceous times.

Tycho's date could be the real indicator of the "big event" that gave origin to the Moon and it ties very well with the writer's theory and it seems to be the last impact recorded so far, then we could find the way to explain that this piece of our planet broke loose just few million years ago, during the early Cretaceous and it was part of an anorthositic craton located in or very close to the South Pole mixed with some upper mantle basaltic rocks.

Coming back to our Earth, this catastrophic event ended with a considerable big piece of the planet taken away at the end of the Jurassic or early Cretaceous.

The huge cavity left in the Pacific Basin disrupted the gravity center moving the rotation axis and therefore the poles of the Earth, at the same time that the new Earth-Moon system was trying to find a common point of gravity to keep going around the sun. This was eventually reached in a point that is now named barycenter (As the Moon does not orbit completely around the center of the Earth and the system wobbles around a common point of gravity located many miles away from the center of Earth).

Those colossal changes must have brought all sorts of havoc on our planet for more than thousand years until all tectonic movements started to slow down and floods and heavy winds smoothen all rough edges on surface getting everything ready to start again.

First we must consider that for a short geological time of few hundred years, all continents were grouped in the Atlantic side of our Earth while the whole Pacific side was partially empty until water and debris from the broken edges started pouring in the big cavity, while the hot magmatic forces of the mantle, trying to compensate for the loss of weight, developed two main sets of fissures that allowed the flow of many billion tons of lava to the surface and from there along the bottom of the basin.

This mixture of lava and water might have released tremendous amounts of all kind of gas and vapors from the Pacific basin contaminating the atmosphere and giving raise to the formation of new islands.

On the Atlantic side, all continents started breaking apart from a huge north-south running fracture developed in what is now the mid Atlantic. It was opening like a zipper from the south pole, on the opposite side of the empty basin, extending upwards as everything else was also moving towards the big hole of the Pacific.

This fracture is still active at present times and is constantly filled by lavas flowing from the magmas in the mantle, forming a mid oceanic ridge which have been steadily growing and widening in a process of expansion towards the Pacific as shown in Fig.8.

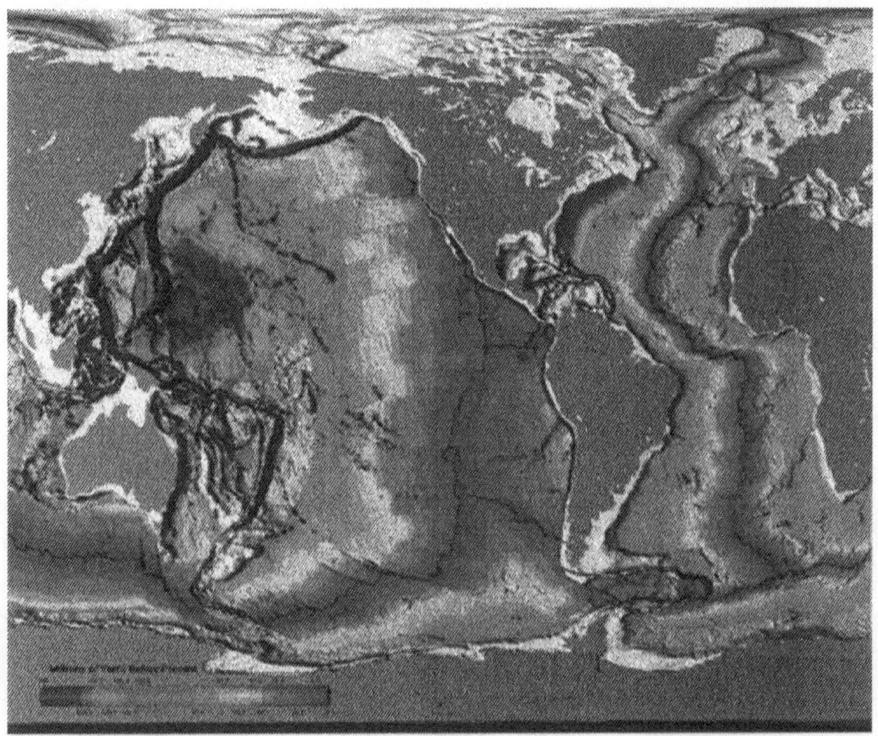

*Fig 8. This map was prepared by NOAA and shows dramatically the movement of the tectonic plates towards the continents. The red colors represent the youngest sediments. Note the bigger "wound" on the Pacific to the west of America and the southwest of Australia compared with the clear "zipper" on the Atlantic side. Australia moved to the present place to compansate for the loss of mass in the Pacific while Antarctica moved above the South Pole..*

On the Pacific there are two main sets of fractures one in the Eastern Pacific and the others to the West of the Americas. These fractures are deeper down on Earth and more active as they have more feeding from the closer magmatic chambers. These ridges spread rapidly but the material do not have the opportunity to pile up in tall heaps. These could be the result of all landmass being contracted on the Pacific side as all the continents meet the opposition of the oceanic plates.

South of New Zealand there is a process of subduction in progress with the creation of a dip sea trench. There is a geological ridge known as the Macquire Ridge forming the boundary between the Australian and the Pacific plates. These plates are moving past each other in opposite directions and in 1989 triggered a massive earthquake of 8.2 magnitude.

That could also explain why all coasts on both sides of the Atlantic are shallow and quiet, opposed to most of the Pacific coasts around the world which are rougher and dip, forming a rim also known as "The ring of fire" along which most of the volcanic activity is concentrated.

Let us again consider the second possibility of our theory, in which we assume the Moon was originated in the Precambrian, then it will be difficult to explain the reason why all continents remained clustered in one side with the empty Pacific ocean on the other for almost 3 billion years.

Also it will be difficult to be explained by the religious, as the bible said that in the third day there were fruit bearing vegetation in land then we should think that there was developed life in the Precambrian…Could be that God changed his mind then too…?

Probably we should not be surprised, after all, the business of creation is a rather complicated one, even for a perfect artist.

If this was the case, then, after the big crash, when the Moon departed from Earth during the early Precambrian, the enormous remaining cavity was filled with water, molten material and debris in the manner already explained, but as the Earth started wobbling about trying to find a new rotation axis the polar regions were moved constantly bringing what it could have been one of the ice ages over the whole Earth. After this, a new era of development could have started with all continents buckled together in one side and a real big open sea on the other for few billion years until separation started to take place on the late Jurassic or early Cretaceous.

The main point of both possibilities of the same theory is that the Moon departed from Earth when our planet was already cold and fully, or almost fully developed.

Before concluding this subject we could work on an example:

Lets us imagine there is a small planet the size of Mars that hit out Earth where the North American continent lies now (…Please God this is just an example…) in the way described when we explained the Moon's orbiting processes.

Then, we will have the North American cratons, the sedimentary sequences of the Rockies and Apalachians plus some mantle rocks(still hot), that incidentally will melt down to form basaltic lavas. This irregular body will keep on breaking and re-melting till all surfaces are rounded in the shape of a planet but, in the mean time all type and sizes of the remaining pieces will keep on falling on the new satellite's surface forming craters of different sizes.

The difference here is that as we also have a sedimentary sequence in the new moon rocks, the presence of water could lead to the development of some kind of metamorphic processes the way we know on Earth…Unless the lack of atmosphere absorbs all the existing water and then, we have a copy of our own Moon…

Now the skeptics might ask why God would do things like that?…Why He did not do it that way from the beginning? Well as God is continuously trying to show man how to do things, he made a demonstration for us to learn what creation and evolution is all about. He does not need thermo-nuclear plants to produce energy, like he does not need a freezer to produce ice. For his work on the universe God always used his own created nature to generate other things and then, he needed the Moon to off-balance the Earth in order to slow its speed of rotation by the creation of a barycenter for the new and unique "planet and a bit "pair in the solar system capable to sustain life.

He needed the lunar tides to bring life from the sea into land. He needed the mid-ocean fractures filled with ridges to keep our planet going as a "continuous operation" using the convection currents to move material from out mantle to keep on building mountains and recycling the Earth's crust and atmosphere. He needed all that in order to get some rest on the seventh day.

# Chapter Seven

## A Can Full of Stars

The subject discussed in the previous chapter bring us to the unavoidable situation that we have to consider theories about the creation of the universe.

On the writer's opinion the person or even better,the team of persons to deal with this problem should be a group of men with the precise brains of Einstein, the clarity of expression of Stephen Hawking, the determination of king David, the wisdom of king Salomon and the Christian moral and conviction of Billy Graham and even then we might be still a bit short in quality because these persons should be thinking very much like God himself.

As the brain matter used here is so hopelessly away from the capacity attributed to any of the above mentioned, we will merely make a small incursion in the science of cosmology and will bring up some of the more accepted theories of creation of the universe with our own conclusions just to show how important is to have God as the source of everything.

Men have been always searching for answers to all mysteries of life and the main one is to find how the universe was created.

The unfortunate experience is that historical facts show that religious leaders were always a strong influence against scientists trying to open new ways to understand nature forcing these men to carry out research under great pressure and in many cases, facing death penalty but even then, much of their work was done with such passion that finally, in many cases, these scientists won their price with the introduction of major changes in the whole world's way of living.

That is the undeniable truth: the natural curiosity of man will prevail forever and his search for answers will bring continuos changes to civilization.

Nowadays, all the existing freedom brought nothing but a big separation between science and religion and the only valuable result of this antagonism is that the zealots of either side look to the opposition with great respect admitting their undeniable importance. Religion and science (Or the other way around for the agnostics) are the two main ingredients that formed modern civilizations, we need both to grow technically and spiritually and the best way to achieve solutions should be to have them together then, it will not be necessary to have wars to develop new ideas like it has happened before.

The race between America and Russia to conquer space was an example of good sportsmanship from which the whole world profited, while W.W.II is the example of new technology learned at very high and painful cost.

We consider that one of the main discoveries for the benefit of all mankind was done on last century by the American astronomer Edwin Hubble, already explained in previous chapters. He found that there are many galaxies besides our own and surprisingly he found that there are all moving away from us and from each other, meaning that the whole universe is expanding.

Using the Doppler shift measurements (These are changes in the apparent wave length of radiation of light or sound, emitted by a moving body, which in this case, is a shift of the light to the red, indicating an increase in distance), Hubble found that the more distant a galaxy is, the faster is moving away from us at about two thirds the speed of light. This became to be known as the Hubble's law and it was calculated that the unit was 50 km per second per megaparcec(A megaparcec is a measure of 3.26 million light years).

Now as an update to this theory there is a couple of interesting points raised by M. Chown in his book Afterglow of Creation. The first is that the red shift in the light of distant galaxies rather than being Doppler shifts alone, is also the result of the time elapsed. Since these wave light from a distant galaxy have been traveling across space to us, the universe have been expanding in size causing the stretching the wavelength of light along with it…Now the question is…Could this factor affect the value of the Hubble's constant…?

The other interesting point is the introduction of the concept of time in these cosmological equations as true time and light are about the only two constant measurements of the universe. (This subject will be discussed in more detail later).

Putting the whole system in reverse we might get the idea that many years in the past all celestial bodies were close together and reversing the process even more from expansion to contraction the whole universe could be brought to a single point at the moment of creation, then we have what Hubble was the first to discover and that is; expansion could have only started after a huge explosion at the moment of creation.

Creation, (or the Big Bang for the pure scientists), started everything including space, time, light, matter, and evolution, and since that first moment everything with the exception of time began expanding.

The application of quantum mechanics brought many new answers to old known problems by introducing the factor time to the study of the particles in the micro-universe, but also brought the Uncertainty Method which proves that nothing but time itself can be exactly defined at one given instant.

Stephen Hawking is the most popular writer on the creation of the universe these days and he sustains that the whole universe could have started at one single minute spot or mini-black hole, or even more he considers that many universes could be developing simultaneously from different mini-blackholes.

He argues that, to describe the whole universe in a simple theory it should be divided in two main parts, the first that tell us how the universe changes with time, and the second to find the initial state of the universe.

The common understanding brings to the conclusion that the universe is governed by definite laws and science is continuously growing by discovering and learning how to apply them.

Another undeniable fact is that those laws are also God's creation and similarly to music and other forms of art, all the mathematics and physics laws are already existent in nature. We just need a musician, an artist or a scientist to pick them up from their neatly hidden places and then, the rest of us will benefit for these findings.

God generally chooses geniuses like Mozart, Beethoven, Newton, Einstein and Hawking to be there at the right time and find the new leads that will be converted into formulas, notes, or other type of symbols that will become new forms of art, discoveries or theories.

These findings are for the benefit of the world to learn, admire and get the benefit of using the products obtained from their application in different industries and fashions that will become part of our daily way of living.

There are no doubts that these leading men and women of our world are God's chosen people.(Including those who considered themselves non-religious, or even more like in the case of Darwin, who was a religious man chased away by the church and by his own feeling of guilt for breaking old traditions).

God used them and like the prophets, they were, are, and will be used again by Him to guide and instruct the rest of the world into better ways of living.

King David knew more than we granted him when he wrote his psalms…

> ."Psalm 19. The heavens declare the glory of God,
> the skies proclaim the work of his hands.
> Day after day they pour forth speech;
> night after night they display knowledge.
> There is no speech or language
> where their voice is not heard.
> Their voice goes out into all the earth
> their words to the end of the world.
> In the heavens he has pitched a tent for the sun,
> which is like a bridegroom
> coming forth from his pavilion
> like a champion rejoicing to run his course
> It rises at one end of the heavens
> and makes its circuit to the other
> nothing is hidden from his heat.
> The law of the Lord is perfect,
> reviving the soul.
> The statutes of the Lord are truthworthy
> making wise the simple.
> The precepts of the Lord are right,
> giving joy to the heart.
> The commands of the Lord are radiant,
> giving light to the eyes.
> The fear of the Lord is pure
> enduring forever…

Now,we all know there are also odd theories, wrong laws and bad music but then, the magic old evolution come to the humanity rescue and does not allow them to survive. It is amazing to see how through human history, all good ideas survive the bad ones, and then again we can find some words from Jesus which we think could be applied to this subject even if there were said for entering the kingdom of God.

> *Matthew 7:15."Watch out for false prophets. They come to you in sheep's clothing but inwardly they are ferocious wolves. 16. By they fruit(ideas) you will recognize them. Do people pick grapes from thornbushes of figs from thistles?.*
>
> *17. Likewise every good tree bears good fruit, but a bad tree bears bad fruit. 18. A good tree cannot bear bad fruit, and a bad tree cannot bear good fruit. 19. Every tree that does not bear good fruit is cut down and thrown into the fire. (Or become extinct)…*

In is popular book A brief History of Time, Hawking explains what some people say."…God, being omnipotent, could have started the universe off any way he wanted. That may be so, but in that case he also could have made it develop in a completely arbitrary way. Yet appears that he chose to make it evolve in a very regular way according to certain laws. It therefore seems equally reasonable to suppose that there are also laws governing the initial state…"

These laws might indeed exists, but might be pretty close to God's "own chest" and it seems logic to conclude that those will be the last thing we will learn before reach heaven ourselves. The fact is that these laws are already written in nature, we just need more wise men to discover them.

Then Hawking continues explaining that the theory of relativity and quantum mechanics should some day unite.."Now if you believe that the universe is not arbitrary, but is governed by definite laws, you ultimate have to combine (all) the partial theories into a complete unified theory that will describe everything in the universe…"

For us believers this is no problem, we know this unified theory exists as king David said."*The law of the Lord is perfect*"…

The Russian physicist, Alexander Friedmann worked on Einstein theory of general relativity and was the first to predict that the universe is expanding. He worked on a model assuming that the universe looks the same in whatever direction you look (Similar than you see light in any direction) and from wherever you are in the universe because at any point in the universe all galaxies move away from each other. He started assuming that If the average density of matter in this model is less than, or equal to a certain critical value then the universe must be spatially infinite and the expansion will go forever but, if the density of the universe is greater than the critical value then the gravitational field produced by the matter curves the universe back into itself and is finite though unbounded, like the surface of an sphere.

The interesting point of this theory is that if we start traveling towards the edge of the universe, (That is assuming we will ever know which way the edge is), we will never reach it as it is continuously getting away at increasing speed.

An analogy could be a house fly deciding to follow the same direction of a passing Boeing 747 at the end of the runway. The further they get from the airport the greater the distance between them. The conclusion could be then, that the universe is not infinite, although it just has unreachable boundaries.

Now, if the Big Bang started as a single explosion, it should be logical to assume that the degree of expansion should be decreasing as each particle gets away from each other and even if we ignore the gravity forces there may be always some kind of matter trying to restrain it. (Some cosmologists think that the so called "dark matter" is the responsible for this to happen).

Hubble and his followers have proved that the further away the stars are, their velocity increases getting close to the speed of light…This is other point to be considered further as our galaxy, (bunched together with few others forming the Local Group) is also getting away from the "central point of the Big Bang" (or of Creation) at an accelerating speed related to expansion.

Einstein reached the conclusion that the universe started from a single event using his famous formula E=m.c sq. From there we establish the close relationship between mass and energy.

Mass could be identified with matter at the beginning of the universe and then if we use a good definition of Timothy Ferris in his book "The Whole Shebang", we say that "matter is frozen energy", those are good words to use for the origin of everything…

A fantastic amount of energy was suddenly converted into matter that gave way to particles or protons that formed hydrogen and became later deuterium which mixed with more hydrogen to form helium and continued the chain till we got oxygen and then, as the bible said, we got water,…the source of almost everything.

We need the water compound to form most of the minerals on Earth and then, to reach the final target, the creation of life and man but, lets forget religion and get more scientific by trying to bring in some basic principles as described by Hawking.

There are four fundamental forces governing the universe presently being investigated in depth on subatomic particles found by quantum mechanics. Their results have been projected (so far not very successfully) to a complete solar system up to a galaxy and everything we know.

As the connection between the micro and macro world still have many loose links we will only try to orient the main pieces of this chain to reach a general, utterly philosophical conclusion.

The smallest expression of science starts with the application of quantum mechanics to subatomic particles moving in and around the nucleus, like particles of light that were originally classified as waves.

Einstein found that both definitions seem to be right so for the sake of simplicity we will say that light is composed of particles moving oscillating in different directions but following general wave lines.

Using quantum mechanics it was found that the movement of these particles could be measured on "spins".

A spin is the way to find what the particle looks like from different directions at a given time. They normally have 0, 1 and 2 spins. A particle of spin 0 is like a dot, it looks the same from any direction, but a particle with spin 1, looks like an arrow, it looks different from all other directions until it turns to the original position 360 degrees. A particle with spin 2 is like a double headed arrow, it looks the same each 180 degrees. But there is still the half spin particle which is the one that have to be turned twice over (360 degrees), to look the same again.

Hawking sustains that the half spin particles make up the "matter" in the universe and particles of spin 0, 1, and 2 give rise to forces between the matter particles.

These matter particles are found to obey to the known Exclusion Principle formulated by W.Pauli in 1925. This principle says that two particles cannot exist in the same state at the same time with the same velocity, meaning they will not collide with each other.

So the four major forces governing our universe are as follow:

1. The first is the gravitational force, which is considered the weakest of all forces, the unit of measurement of this force is the Graviton (Spin 2. This particle was calculated but not actually discovered). This force always attracts, never repels like the electromagnetic force and works over extremely long distances. The Earth and sun are controlled by gravitational force.

2. Electromagnetic force is the strongest. Each atom have the nucleus and electrons attracted by electromagnetic forces

3. The weak nuclear force (radioactivity) acts on matter of particles with half spin and not 0,1 and 2.

4. The strong nuclear force hold the quarks (The fundamental particles) inside the atom together. These are parts that form the neutron and electrons. The unit is the gluon which have a spin of 1. This fundamental force have a special property called confinement as it binds particles (quarks)into combinations of different "colors or groups", so then we have that each proton and neutron are formed by a combination of three quarks of different color each and altogether match into a set of six.(For more details Hawking's book should be read).

To make thing even more complicated we have the antiparticle. For every type of particle there exists an antiparticle with opposite properties such as electrical charges (Electron(-) and antielectron or positron(+). The meson is a union of particle and antiparticle of the same color. It is very unstable because quark and antiquark can annihilate each other producing electrons or other particles.

All this bring us to the work done by Tsung Doo Lee, Cheng Niug Yeng and later Chien Shiung Wu (believe it, they are all American's Nobel price winners). They worked on the combined symmetry in the weak force. If a particle is equally replaced by an antiparticle it should be the mirror image but they found that this is not the case as the electrons give more in one direction or the other. The importance of this discovery is that it could help to understand the idea of the expansion of the universe.

These findings are brought together with the intention of looking for a connection leading to understand the formation and composition of the "dark matter" which is a mysterious component supposed to hold the largest mass in the universe.

This could also help us try to understand a bit better the mind of God, like we said at the beginning he made this world "almost" perfect leaving small irregularities as "open ends" to be corrected with new developments created by man or himself. Everything inside the universe is finite and dynamic, meaning it has a beginning, a growth period to maturity and finally death but what about the universe itself…?

If we are concerned about our world ending in the next few million years there is a good chance that another new world might be developing at this very moment right behind us, closer to the center of the universe…

In 1965 Arno Penzias and Robert Wilson, two American physicists working for Bells Telephone Lab in New Jersey, picked up extra noise in every direction they moved their microwave detector. After careful examinations they found out that this radiation, originated by the Big Bang, come from outside the atmosphere and even more beyond our galaxy as detection was made regardless of the Earth position around the sun and of any direction they pointed the detector. This discovery is comparable in importance with the work of Hubble and was the main factor indicating there is an expansion of the universe. Measurements done later from instruments located on special planes and more recently in the COBE Spacecraft (Cosmic Background Explorer Satellite) confirmed the importance of these microwave radiation.

Our view is that this cosmic radiation sphere is a clear indication of a single starting point for all creation as the cosmic microwave background is nothing but the initial edge of the universe

Stephen Weinberg describes the Big Bang explosion in is book "The First Three Minutes" as an explosion not like those familiar on Earth, starting from a definite end and spread out, but this explosion occurred simultaneously everywhere, filling all space from the beginning, with every particle rushing apart from any other, and in agreement with the second model of Friedman he thinks that all space means all of an infinite universe which moves back into itself like the surface of an sphere.

Our point is; if we assume that expansion was originated in a single point, with single particles splitting at fantastic speed while getting away and expanding equally in all directions from the center, like an immense fantastically fast growing sphere. This should bring the universe to have an inner "core", which could be void or filled with unknown matter in contact with a :"mantle", comparable to the one in our Earth's, and an edge where everything becomes energy. Lets imagine an immeasurably big coconut with the inner part filled with juice that form the "core", the white pulp is the "mantle" where all the matter gives way to the formation of the galaxies we live in and the shell is the edge where we all going to end up as energy.

Now this "coconut" is alive and somewhere in the inner mantle, close to the boundary with the core, there are concentration points or nucleus that give formation to all galaxies which start to grow while traveling outwards, reaching maturity and later death when reaching a critical speed near or at the edge of the universe.

At this critical point, all mass disintegrates into single particles of matter (Dark or bright), to become "frozen energy", moving along or near the edge in the search of an access zone where to re-enter the universe (like the convection currents in our mantle but here could be represented by a black hole). Then this energy starts traveling inwards to a point where there are repelled by the nucleus or core where eventually will concentrate around some central point or nucleus near the boundary of the core to re-start the process of galaxy formation.

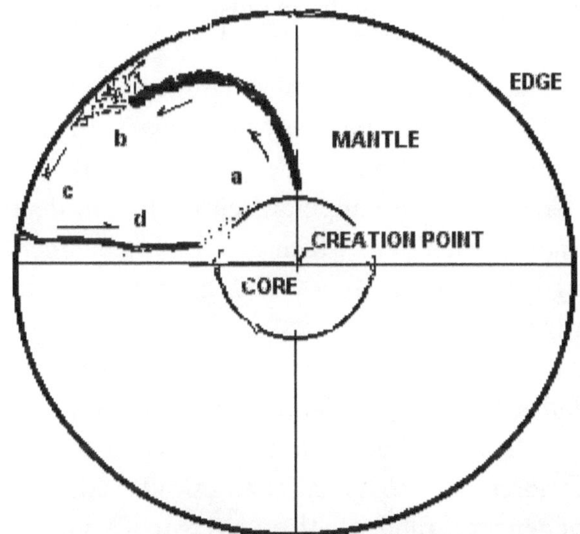


*Fig.9 Evolution of the universe. Point (a) is where a galaxy is getting formed , growing and maturing till reach point (b) when it becomes energy that moves along the edge (c) till reach a point where to enter again toward point (d) where galaxies start regenerating around concentrated matter.*

This looks very much like the process of regeneration we constantly have on Earth(See figure 6 in page 45). This concept could show once again that God have a perfect system that repeats and regenerate itself in different dimensions. This means that the whole chain in the universe formation could be started considering an atom with its strong nuclear forces, surrounded by electrons that join others to form molecules that will become matter which are part of our life and our solar system, with the sun also having similar nuclear forces to control its planets around and all together traveling around the center of our galaxy which, with others, form part of God's universe.

Another way to explain this process would be to assume that the Milky Way Galaxy containing our solar system could be somewhere in the middle of the "mantle" of the universe giving us the impression that we are located in the center of the explosion as everything is getting away from us.

This concept of a growing universal sphere seem to be is a way similar to the fire ball as described by George Gamow in 1948 and followed up by R.Alpher and R Hermann.

They considered that at a temperature of a billion degrees, matter produces internal gamma rays which is a highly energetic radiation with a wavelength far shorter than visible light making the early universe a brilliantly bright fireball.

This fireball could contain large number of free electrons which are particularly good at absorbing radiation and redirecting or scattering it.

When two electrons collide with each other, the one with more energy generally loses some to the other and similar effect occur when a photon of higher energy collides with an electron. (A similar concept to the one already explained about the work done by Doo Lee, Nieg Yen and Shing Wu, finding some lack in symmetry of particles and proving the universe expansion.)

In the Big Bang fireball, as the universe was expanding the interaction between photons and electrons were also happening at an increasing speed so energy was spreaded evenly throughout space maintaining the "whole system" in equilibrium. This state of equilibrium does not mean that the energy of each individual particle is unchanging as we explained before all particles continue to be involved is this permanent exchange of energy.

After a fraction of a second from the Big Bang we had only photons, electrons, neutrinos and their antiparticles with the formation of few mesons, at a temperature of about 100,000 million degrees Kelvin.

This fireball kept growing continuously as an sphere with an apparently unreachable edge which is also continuously growing. We assume that the left over of this first moment in the universe is the microwave background halo that Penzia and Wilson discovered in 1965.

Looking inside the matter of the immense sphere that forms our universe, our galaxy could be visualized as a small spot somewhere halfway in any direction.

If we apply the factor time to this process as from the beginning, 15 billion years ago, few seconds after creation the universe was very small and dense but expanding very fast. The velocity of each proto-star increased getting away from the center and as galaxies were forming and getting further away from each other inside this "soup" of cosmic matter, which is also moving towards the edge at an increasing speed, it might be a moment when the speed of a galaxy becomes critical and disintegrates becoming "frozen energy".

It seems likely that all this generated energy will be running along the edge till finding a black hole or another mean to get their way back into the system in a similar way than the convection currents inside our Earth carry out the heat that keep forming our mountains.

This would make our universe a closed circuit comparable to the way we have in Earth already explained in chapter six, where we have a inner core of heavy metals generating the heat to move the materials of the mantle by convection currents towards the crust where we live in.

The Universe might have a core, where everything started 15 billion years ago, that will be the one of the last mysteries of creation to be solved. Surrounding it, is the "mantle", the spherical ring where all matter concentrates giving place to the re-birth of the galaxies. There is something on the Bible about this too:

*Ecclesiastes 3:14. I know that everything God does will endure forever; nothing can be added to it and nothing taken from it.God does it,so men will revere him.*

> *15. Whatever is has already been,*
> *and what will be has been before*
> *and God will call the past to account.*

The model described before seem to be partly in agreement with the first model of Friedman where he thinks that expansion would go forever but the difference in our view is that the universe has an edge and therefore is not infinite. This edge is growing by the minute and the big question for all this might be, when all will end. We assume that our galaxy is not in the center of the universe, or better said we do not know where we are in the universe,(and this could be a link to use the complexity of quantum mechanics, or at least the Uncertainty Principle), but we know that our sun, like many other stars getting away from us, must also be the "getting away" star from some other starts and galaxies closer to the center and we also know that our velocity towards the edge is increasing by the minute then by finding out the rate of increasing speed of our galaxy we might be able to find out how long we have to go before we become particles of energy…

If we go back to the bible and start again at the real beginning we read…"*In the beginning God created the heavens and the earth. Now the earth was formless and empty…*

There is no word about the universe, only the little earth, a tiny spot in the middle of the Milky Way galaxy in the immensity of the universe. Could be heavens a equivalent term for universe?…It seems is not, as in Psalm 19 king David said that…" *the sun rises at one end of heavens and makes its circuit to the other*"…Then we could think that the words of the Lord was only referring to our solar system which we willingly could extend to our galaxy but,…could it be that the universe was already existent when our galaxy was born…?

God promises were (Genesis 8)

22 *As long as earth endures,*
*seedtime and harvest,*
*cold and heat,*
*summer and winter*

> *day and night*
> *will never cease*

and in Genesis 9,…*14. Whenever I bring clouds over the earth and the rainbow appears in the clouds, 15 I will remember my covenant between me and you and all the living creatures of every kind…*

We now know that the velocity of our galaxy will keep increasing till it reaches the critical speed, probably far before reaching the speed of light. Then, the whole galaxy will be converted in energy but, what will be happening few years before?

> *…Revelation 8…7. The first angel sounded his trumpet, and there came hail and fire mixed with blood, and it was hurled down upon the earth. A third of the earth was burned up, a third of the trees were burned up, and all the green grass were burned up.*
>
> *8. The second angel sounded his trumpet and something like a huge mountain all ablaze, was thrown into the sea…*
>
> *10. The third angel sounded his trumpet, and a great star, blazing like a torch, fell from the sky…*
>
> *…12. The fourth angel sounded his trumpet, and a third of the sun was struck, a third of the moon, and a third of the stars, so that a third of them turned dark…*

It sounds like we will have enough warnings many years before everything happens…

# Chapter Eight

## Moving Through Space

Now that we opened this can of stars, we are compelled to try to see where we really are situated in the universe to be able to explain the miracle of our existence. We are going to get deeper in the astronomy field hoping not to step over some specialist's toes in our search.

The whole universe structure must be an extremely complicated mechanism in which immense amount of energy is produced, consumed and re-generated during the whole elaborated process of formation, rotation, orbiting and explosion of different kind of stars.

We know that our Earth orbits the sun at a velocity of 30 Km per second, the sun orbits the galactic center at 250 Km per second and the galaxy traces an orbit with other galaxies of the called Local Group and this group share a bulk motion of over 630 Km. per second.

All these movements result in the phenomenal formation of energy centers named Jetstars, (A process where radio emission are created and electrons are accelerated to nearly the speed of light moving through magnetic fields like jets and shot out in opposite directions from the center of a galaxy). When these jets reach the outside rarefied gas of the intergalactic space they loose the directional motion becoming huge radio-emitting gas clouds or Spinars (Large spinning stars that attract gas into their surface getting some to spin around being heated up and later expelled as jets due to centrifugal forces.) or many other form of stars well known in the astronomy books that will br not described here for the sake of simplicity but all are proved to be other sources of energy

Fig.10. This is COBE image of the Milky Way. It has a collection of about a hundred billion stars one of them our own sun.
(After Microwave Anisotropy Probe.)

Our solar system is sharing the Milky Way galaxy with at least 100 billion other stars and their planets. The Milky Way have a Galactic Center around which we all orbit about. It also have a luminous band spanning all around the sky along a symmetry plane known as the Galactic Equator.

Astronomers calculated a total mass between 750 billion and one trillion solar masses and a diameter about 100,000 light years. Our Milky Way is considered to be a giant galaxy but is not alone. It belongs to a "Local Group" formed by three large and over 30 small galaxies. The Milky Way is the second largest after Andromeda.

Our solar system is situated within the outer regions of the galaxy, about 28,000 light years from the Galactic Center and about 20 light years above the equatorial symmetry plane.

Our solar system is orbiting the Galactic Center on a nearly circular orbit. We are moving at about 250 km/sec, and we need about 220 million years to complete one orbit,(Our preliminary correlation of geological events with the book of Genesis indicated a length of 190 million years to the fourth day when the Moon was created. Could that be the unit of measurement for one day?), This means that the Solar System has orbited the Galactic Center about sixty eight times since the moment of creation and the Earth have completed only about twenty since its own birth 4.6 billion years ago.

Analyzing the COBE image of our Milky Way, astronomers concluded that the Galaxy has three major components.

> 1.*A thin disk consisting of young and intermediate age stars This disk has spiral arms which are the regions of active star formation.*
> 2. *A bar of older stars seen at the central part of the picture shown by COBE.*
> 3. *An extended dark halo whose composition is unknown but its existence is proved by the gravitational pull on the visible matter.*

This halo has also been described as an envelope that consist of the invisible "Dark matter", the mysterious main component of the universe with a suspected mass of more than 80% of the total of the whole universe.

If this dark matter is filling all empty space in the universe then we must assume that its particles are also moving as a result of expansion.

Assuming that the Milky Way is moving at greater speed than the dark matter it is to be expected to form an halo similar to a plane going through air.

T. Ferris explains there is a hot spot in the Cosmic Microwave Background caused by our galaxy motion in one direction relative to the rest of the frame of the universe, with a corresponding cold spot on the opposite side of the sky. This could give us a good idea of the direction in which our galaxy is moving towards the edge. On the writer's opinion this might be another case of "friction" between at least, two different kind of matter or the same matter with different density and this friction might liberate a large amount of energy to contribute with the continue re-generation of the whole universe.

Other explanation for the existence of the envelope of dark matter around the galaxies is that it gives protection to the stars inside against the tidal forces created by the gravity forces of other galaxies.

As it is also known that there is dark matter filling all the space inside a galaxy, the envelope might be a concentration of "darker matter."

Everything in our universe is closely related, from the smallest particles to the biggest stars and every event is connected within a period of time from birth to growth, maturity and death.

We know now quite well the structure of an atom, and even if scientists keep on finding new particles we have basically a nucleus filled by smaller particles and around it there are few orbits where electrons move like micro-planets around its sun. This atoms blend together to form molecules that give the form of all existing matter from dust to the most intelligent forms of creation. All being part of the planet Earth, consistent of an inner core filled with heavy metals, surrounded by a mantle where its components keep on re-melting and move to the surface to form mountains and the continents we live on. This Earth is part of a solar system consisting of a nucleus named sun, composed mainly of helium and other gasses with planets moving in different orbits like the electrons around a nucleus. Our solar system orbits the galaxy around a center which is apparently empty, but following a well defined orbit and the galaxy itself forms part of a group orbiting around a center while moving together in some other direction.

All these movements are relative to each other and all seem to move away from the real direction towards the edge, meaning that the life of our Galaxy might be extended few more million years but, there is one true direction in which time is absolute, from the beginning through all the periods and towards the end of the universe at the end of time.

This complicated pattern in the movement of stars with planets inside galaxies at a given time could be compared with the subatomic particles used in quantum mechanics and derived new theories. The huge difference would be in the time units to be used as the Uncertainty Principle is likely to be applicable considering our particle looking Earth compared to the rest of the universe.

It seems that in our universe everything is almost perfect, and there are repetitive movements following well defined patterns corrected and compensated by natural forces such as the strong nuclear forces, the weak forces, the electromagnetic and gravity forces

These facts bring us once again to think that the whole universe structure have a similar formation, with a nucleus of unknown composition, a "mantle" where the life of the universe develops and an edge where everything ends and becomes regenerated.

The continuous addition of dark matter to fill the growing universe could be partly produced by galaxies exceeding the critical speed and reaching the breaking point at the edge. This huge amount of matter converted in infinitesimally small particles such as neutrinos which has only mass but no electrical charges, might travel along the surface of the edge like the ocean plates move towards the continents, until they found an open spot (or a Black Hole?) or a weak zone where to move back toward the center bringing new life to the inner mantle where new grow points are formed to give origin to new galaxies.

We now know how fast we are orbiting around the Galactic Center and we also know that our galaxy is moving at an accelerating speed of over 600 kilometers per second, but,…Where are we going at such speed together with our bunch of neighboring galaxies?…And, where are we coming from?…Do we know how long we have until our galaxy get the critical speed and we all become particles of energy…?

As science keeps providing continuous answers to many of these questions, it seems certain that it will not take long to find which way to head our spacecraft when the moment comes.

# Chapter Nine

## The Expanding Universe

The last chapter brought us to the essence of a new theory which cannot be claimed as an original idea but as a combination of many explained for different scientists after long time of research

We must start by believing that the Big Bang began as a singular event from a central point that we believers call God and the agnostic scientists could call a single central point from where everything started.

The initial components of "everything" were space, time, light and matter. Meaning that as time started running, the first particles formed were photons, electrons, neutrinos, bossons and all the vast arrangements that gave shape to the first atoms and as the bible said, at that moment, all matter was in a state of chaos. This matter was contained in a extremely hot ball of fire just as Gamow and his team explained, until water as highly compressed gas came into being and then clouds of super heated gases started breaking away from each other growing at incredibly fast rate into different nuclear points that would become primal galaxies getting away from each other like eggs expelled from the mother fish, being fertilized with many stars which would eventually, give birth to planets like the one we are living on.

Now, if expansion started taking place from the same moment of the Big Bang as a huge atomic explosion going in all directions from an original giant ball of fire and if each particle of matter continued moving away from each other filling all the space like a huge growing sphere, all components of the universe are also getting away from the original center which must have matter coming from somewhere to fill the empty space left by the moving galaxies.

Hubble proved that the further galaxies are, the fastest they move away reaching speeds close to the speed of light. This might be the limit we want to measure before reaching to the destruction of an over-ripe galaxy.

In the subatomic world of quantum mechanics mass and energy change unceasingly into each other in such a way that physicist use energy unit to measure mass then we can assume that if a galaxy disintegrates into subparticles it might become energy.

Then, there must be a way for this energy to be reabsorbed into the universe dark matter which is the "soup" where all galaxies wander through in their way to the edge.

A great achievement of the science was to find that the rate of expansion of the universe can be measured using the Doppler effect and results indicate that is expanding by between 5 and 10 percent every thousand million years.

As the galaxies are moving at increasing speed toward the edge of the huge sphere we could assume that the inner part of the universe becomes depleted of matter that must be replaced also at a fast rate.

All this moving around of colossal masses must create some kind of convection currents in the dark matter similar to the currents we have in the mantle of our Earth which keep our planet alive with generation of heat in depth for the re-generation of magmas and further mountain building.

Apart of the unimaginably difference in size, the convection currents inside the universe must have the dark matter as responsible for the formation of new galaxies after completing a cycle inside this close system. This system would be a singular, close, self contained, continuous mechanism in continuous evolution which could only be considered God's creation.

Hawking explains that our galaxy, similarly to others, is growing over a higher concentration of "dark matter". This concentration of "dark matter" could be the result of expanded concentration points from the inner part of the universe that were the nucleus of formation of new "baby galaxies".

Workers on the project COBE found the existence of ripples in the cosmic background radiation, which is the witness of the moment of creation. They noticed that some parts of the sky are slightly brighter and more dense areas that grow continuously as the universe expands.

This bring us to a very important point against the theory on the co-existence of few different universes…We know now that the cosmic radiation is affecting equally our whole universe,…Then,…should we have more universes growing around like bubbles we should also be able to pick up some overlapping of micro wave radiation from other creations…

It seems more reasonable to think that there is only one universe to be considered, or at least, to think that we live in one not affected by others.

The ideas of galaxies formation are not new as Herman Bondi and Thomas Gold from Austria together with Fred Hoyle from England were talking about galaxies moving away from each other with new galaxies forming in the gaps in between in 1948.

Now as part of all this universal ensemble, our solar system is a member of a galaxy that is also getting away towards the edge and we are also getting away from other galaxies in the inner universe at faster speed, then our question should be…How fast is our galaxy moving?…Hawking explains that it is possible to determine very accurately the present rate of expansion by measuring the velocities at which other galaxies are moving away from us so we can assume it will be possible to calculate our own galaxy speed. This is very important as it would help us to find how long we have until we reach the critical speed.

We all know and the bible said it, that the world will reach to an end.

God told us so and he also told us we are like him as he gave us the brain to learn to find solutions for all our world problems. He also gave us the power to discern and choose the right way to do things and for that he also allowed the wrong ways to exist, which most of the times seem to be the easiest.

If we learn how fast we are moving we will learn how long we have to live. My guess is that it might be still few millions years to go as we are about in the middle of the seventh day…Or are we not?…

A good thing would be that once we get to know there is only few hundred years to go, then the best time will start on Earth for everybody as many people will stop building up empires to be remembered by future generations as there will be not many generations to go and then, greed might cease to exist, the huge monopolies will have nothing to grow to and everybody in the world will unite to build a fleet of "Starships Enterprises" that will carry a selected crew of Earth citizens into a new world to be discovered in the inner growing universe…

But we still have to find which way to direct our ships…

By that time our descendants might be closer to God, they will dress in special protected suits of bright colors, a pilot will be wearing the Air Force emblem of an eagle on his chest, on his right upper-arm will be wearing the emblem of a lion from his squadron and on his left upper-arm the emblem of an ox from his homeland. They will move on land using small jet propellers attached to their backs and will have small stabilizers looking like wings to controls their moves. They will land from an special helicopter-module that will be kept in space controlled by another pilot and they will be looking into a new more primitive world somewhere in the inner universe to find a place where humans could live…

*…I looked and I saw a windstorm coming our of the north—an immense cloud with flashing lighting and surrounded by brilliant light. The center of the fire looked like glowing metal, and in the fire was what looked like four living creatures. In appearance their form was that of a man, but each of them had four faces and four wings…All for of them had faces and wings…Their faces looked like this: Each of the four had the face of a man, and on the right side each had the face of a lion, and on the left the face of an ox; each had also the face of an eagle…*

*…The appearance of the living creatures was like burning coals of fire or like torches. Fire moved back and forth among the creatures…The creatures moved back and forth like flashes of lighting…*

*…Spread out above the heads of the living creatures was what it looked like an expanse, sparkling like ice and awesome…*

*…When the creatures moved, I heard the sound of their wings, like the roar of rushing waters, like the voice of the Almighty, like the tumult of an army…*(Extracts from the Book of Ezequiel1:4 to 24. There is lots more for who would care to read further)

Could this mean that God has already send some people back from the edge zone…?…

# Chapter Ten

## World Evolution

Now is the time to think about all this processes and admit how small man is and how insignificant our beloved little Earth is, compared to the rest of God creation.

We have now enough evidence to prove that since the starting of the world history the man's brain evolved with increasing speed from one generation to the next and time will come when many more answers will be found about the mysteries of creation but, like the fly trying to follow the Boeing 747, God's ideas will keep getting further ahead of us.

Some skeptical will ask why we want to know that much, and many religious groups will argue that this matter should be let alone as are all God's secrets and should be kept guarded by religion.

These two groups might agree in asking…What help can that bring to humanity?…Well, the best answer to that is with some more questions…Why men climb difficult mountains? Why men want to go out to space?…Why Columbus started his trip to discover America?…Why scientists try to fight all diseases?…Why man is trying to find new and faster way of transport? Why people are trying to improve the speed and capacity of their computer and find new ways of entertainment?…Why they look for new places to visit?…new supermarkets to shop on,…new houses to live in?…new clothes to wear?…

That is an important attribute of human nature and there will be no laws or religion that will stop man from asking all questions about the universe. Man is inquisitive by nature, God made him like that. He created man that way and made him to His resemblance, not because God looks human, but because we have a brain and a spirit (or mind, or soul) attached to it that make us capable of conducting our own little creations, improvements and destruction in the world we live in.

He made the best, the almost perfect machine, our body, (Many times abused by their own users by pumping in excessive food, liquids or unnecessary drugs). This machine was specially designed to carry around our brain and soul, our personal treasure, our God's legacy. That was the main function that God had in mind for our bodies, other uses were secondary but were to experience life as God asked us to live, growing in knowledge and reproducing in order to transmit our knowledge to the next generation for them to continue one step ahead in the process of human evolution.

So…why there is an argument between evolutionists and creationists?…God also created evolution probably at the same moment when he created time. We know now that there is a continue change in the whole universe from one moment to the next in time. Quantum mechanics is continuously trying to prove that everything changes in time from one fraction of a second to the next and we explained how our planet is also moving each second to a new unknown position inside our galaxy in the way to the edge.

History has proved that Darwin theory is basically right. He sustained that in any population of self reproducing organisms there will be variations in the genetic material resulting in changes in the upbringing of new individuals and of those the strongest will prevail over the others making the changes a new rule of life on that community. The better able to handle life situations will be the more likely to survive and reproduce and their pattern of behavior will dominate.

Comparatively, the greatest civilizations on Earth dominated others, imposing their way of living but our history shows that in all cases these big empires became wicked by sin and collapsed even when they were more advanced in knowledge and more powerful than other slower growing civilizations. This seems a step back in evolution but is not so, because these lower civilizations that eventually took over control, were not inferior, just a bit behind, probably because they had higher moral principles and less aggressiveness which is the big difference with any other living species. In this human evolution, the slower growing civilizations absorbed lots of useful learnings from the invaders that were adapted to the way of living. (The bible give many examples of this happening in Old Testament times). In that way, civilization keeps always going forwards

As I write these notes I can see the snow falling on my garden and to me this is proof of God's existence. No man can stop the snow, the rain, the sunshine and all God given blessings that we call nature.

In the book of Genesis, at the end of each day the bible said, "and God saw that it was good"…like an artist critically looking at his masterpiece checking whether everything balance well. God is the greatest artist of the universe and when he saw it was good he let it be and then, instructed his writers to put it on the bible…but…what about things that he created and after did not like?…Yes, as a the creator of everything in universe and under the sun, He probably noticed some of his products were getting a bit out of shape…Like the dinosaurs…

He was doing all the work, adapting parts of his own creation for the future use of His final product,…man,…God did everything on our galaxy, our solar system and our Earth, for man, and specially on our planet, for us to rule everything on his behalf.

So, in the case of the dinosaurs, He decided there were too big and too stupid to live and to be useful to man, then he quickly erased them like He did after with many other things he did not like.

He did not explain that in the bible because God knew that all those primitive men, living in permanent fear, were constantly drifting away from Him by inventing and worshipping idols shaped as gods and they did not deserve further explanation.

Going back to geological times we know for certain that our Earth had few Ice Ages at the end of the Precambrian just before multicelular organized life started to populate the planet.

Some modern scientists are trying to prove that the whole Earth was covered by ice in the Neoproterozoic (Late Precambrian) as they found clear evidence of glaciar existence near the equator during that time. Paleomagnetic measurement helped to establish that there were no polar changes on Earth at that time and similar "glaciar in-prints" were also found in Australia, Kenya, Europe and North-America between 900 to 544 million years ago towards the end of the Precambrian.

Much later in the Pliocene and Pleistocene, just before man started to rule the world, there were more glaciar ages recorded in different times for the northern and southern hemisphere, indicating some instability of the Earth Moon system in relation to the Sun.

These periods lasted few hundred years all around our planet and were deliberately not mentioned in chapter 2 in order to keep the fluidity of our thoughts, the same as the bible did not mention it probably to make everybody's life easier.

These events seem to be of a larger scale than the flood but as they occurred when probably the Earth was still barren and there were no people involved (Or at least no people worth to be mentioned by God) the bible does not seem to talk about these things…or does it?

Lets go back to Genesis 1:6…"*And God said, "Let there be an expanse between the waters to separate water from water." 7 So God made the expanse and separated the water under the expanse from the water above it. And it was so. 8 God called the expanse "sky." And there was evening, and there was morning—the second day.*

This means that there was only sky with water above which could be taken to be clouds and water all below over the Earth and frozen water becomes ice…?

Joe Kirschvink in is paper named "Snowball Earth" describes the world at Neoproterozoic time as…" entirely sheathed in ice, with the oceans smothered by a freezing white blanket more than a kilometer thick. Nothing moves and there are no clouds, save perhaps a handful of high wispy streaks made by frozen crystals of carbon dioxide and temperatures about-40 degrees Celsius."

Kirschvinik is a specialist in paleomagnetism working for Caltech in Adelaide, Australia. Rock samples taken from a Neoproterozoic glaciar deposit revealed that there were folded and slumped while there were still soft, yet they bore a magnetic imprint that had been applied before the folding proving that the magnetism was the same age as the rock and that the Ice age was the same at that latitude and also the tropics were iced over at that time.

At the same time other geologists found evidence of Neoproterozoic glaciar deposits in Namibia and later other areas of the world including areas around the equator.

Could those events be a later result after the Earth was grazed by another planet getting our Moon into orbit…?…This could be in agreement with the age determination of the rock samples analyzed from the Moon. If that was the case then, we will have to accept that the continental drift and the consequent plate tectonic are the result of other major event on Earth during the Mesozoic not related to the origin of the Moon.

Now the skeptics ask. Why should God do all that?…Why change his mind if He is perfect?…Well we find an explanation to that…He knows how to make things perfect but as we said at the beginning He did not want to make our world perfect because He wanted man to take over and show Him that could become clever enough to fix the few imperfections that God put in the world for us when we reach close to His level.

So let assume that God froze the world on the second (Or fourth?) day of his creation while he was thinking what to do next, or because he needed that to happen in order to build the world the way He wanted, and then, if we go back to the Bible we read.

> ...9 *And God said, "Let the water under the sky be gathered to one place, and let dry ground appear." And it was so. 10 God called the dry ground "land" and the gathered waters he called "seas". And God saw it was good...*

Now these flashes of creation mentioned in the Bible describe some immense events that changed the whole world.

We all know that our Earth is alive and even on those early days of formation, our planet already have the central core rich in heavy metals at super hot temperatures and super high pressure, surrounded by the mantle with all the hot magmatic bodies looking for a way to escape to surface. Then, a small crack was only needed to release pressure and form a funnel to allow all gas and liquid to flow upwards blasting the ice away and worming the atmosphere.

The specialists view is that the ice of the oceans melted first, allowing water vapor back into the atmosphere. These were the most potent greenhouse gases there could be, that added to the existent $CO_2$ could reach temperatures "rocketing" to 40 or 50 degrees C, so according to D. Schrag, a geochemist of Harvard,(New Scientist 11.6.99) in a few hundred years you went from the coldest climate the Earth has ever experienced to the warmest ever.

Also these changes in temperature with the soaring levels of $CO_2$ would form torrential acidic rains and "hyperhurricanes", accompanied by other colossal event that helped reshaping our Earth.

Now, all these scientists described these events as "unexplained mysteries" (New Scientist 11.6.99) and adding to these their Namibia data, it looks there were at least two cycles of what they called Snowball Earth followed by acidic hothouse. Other evidence around Earth could prove the existence of up to five different cycles.

They also explain that life on Earth started earliest with simple, single-celled creatures such as bacteria and algae. The most important after the last "snowball", was Ediacara, an strange organism that might be the first multicelled being and after Ediacara came the traditional known explosion of the Cambrian with most of the species known today.

So science comes to our help to understand why God might have done all this strange things in Earth before Adam. God was using all components on Earth to reshape the planet to form the continents and He needed a pleasant atmosphere and enough humidity to allow things to grow.

A geologist from Harvard P.Hoffman (New Scientist 11.6.99) continued explaining that this snowball might have been triggered after the disintegration of Rodinia, the massive supercontinent that used to encompass all the world's land. He thinks that till then, the continent interior would have been arid as the oceans were too far away to provide moisture for rain, but after the break-up rain would have fallen plentiful across the landmasses. With more rain, comes more chemical weathering, producing more $CO_2$, taken out of the atmosphere and locked up in carbonates which produces more cooling…and a snowball.

All these theories, better explained on different published papers, bring us to the conclusion that we might owe our existence to these hot and cold flushes that helped to the formation of thick sedimentary rocks which were part of the Earth growth. To this we can add that these thick sedimentary layers created enough pressure to develop cracks on the crust below which allowed the magmatic solutions to come close and in some place into the Earth surface as basaltic and andesitic lavas.

So if we ask again…Why God do such "unfinished things" like that if he if perfect why he did not do a perfect world?

…Well we all know that He created paradise and we proved not to be fit to live in it…But,…there is a degree of almost perfection everywhere and God showed us that everything we need to reach perfection is already found in nature, our planet is alive, healthy and capable of producing everything we need if we learn how to use it. He gave us coal for heating our houses and when we grew to learn that gas was better it was also there for the take, then we learned more and started using hydro-electric plants and later nuclear ones and perhaps some day we will learn to store the already made energy from the sky, hidden in the clouds and in thunder-storms. These would produce enough electric power to keep all the big cities of the world lighted and warm.

> 11. *Then God said, "let the land produce vegetation: seed-bearing plants and trees on the land that bear fruit with seed in it, according to their various kinds." And it was so.* 12 *The land produced vegetation: plants bearing seed according to their kinds and trees bearing fruit with seed in it according to their kinds. And God saw it was good.* 13 *And there was evening, and there was morning—The third day.*

Using some of the scientifical data described, it could be said that after the "snowballs times" and all those colossal Earth events, God allowed life to start on Earth on the Cambrian.

This is not the last time that God used cold weather to fix our Earth, just before the time of man creation we had the traditionally known Ice Ages. There is a difference in the time of these ages around the world, It happened first in the southern hemisphere, in the Miocene (23 to 5 million years ago), while in the northern hemisphere was in the Pliocene (5 to 1.8 million years ago). Could that be that God was preserving something for the upcoming of man? If that is so then these Ice Ages happened at the end of the sixth day…

W.Lee Stokes (The Essential Earth History) claims that the ice age was far from simultaneous everywhere and the latest glaciation periods started about 5 million years ago in the north Atlantic of northern hemisphere and about 13 million years ago in Antarctica in the southern hemisphere. This estimates are based on data obtained from deep sea sediments.

The Pleistocene followed with notable changes involving the creation of new species and even new genera including the first australopithecine species that were going to become men.

All these geological events proved mainly by studying the stratigraphy of our planet have been also described in the bible in a different way. Our co-relation might not be accurate and we admit there are further studies to be done by revising the sedimentary sequences as well as different passages of the Bible. Going back to the more exact sciences there is another example of this almost perfection in the conception of the universe these are given by the six numbers of Rees.

According to Martin Rees, a British astronomer from Cambridge, there are six numbers that rule the whole universe which are:

1) The number that describes the forces that binds atomic nuclei. This number shows that a nucleus of helium weights 99,3% as much as 2 protons and 2 neutrons that fuse to make it and the remaining 0.7 % is related as heat, which is the fuel that powers the sun, where the hydrogen gas at its core convert 0.7% of its mass into energy when it fuses into helium.

Should this number be reduced to 0.6%, then a proton could not bind to a neutron and the universe would be only of hydrogen, no chemistry, no evolution to life in it. Should it been increased to 0.8%, then fusion would be so ready and rapid that no hydrogen would have survived the Big Bang.

2) Other is the number that measures strength of forces that hold together two protons, divided by the force of gravity between

them. (This is a very big number, a one, followed of 38 zeros), this means that gravity is rather weaker than the inter atomic attraction. If this number was smaller only a short lived universe could exist.

Paul Dirac noticed this fact in 1937 and he compared it with the ratio between the time it takes light to cross the universe and the time it takes light to cross an electron which is a one followed by 37 zeros. This closeness helped Dirac to think about a relationship between these two numbers, indicating a need of finding a way to equalize them.

Knowing about the universe expansion, that was a new idea on those days, it would mean that the time for light to cross the universe will increase as the universe gets older. Then Dirac suggested that the electrical to gravitational forces ratio would also have to increase with time. His conclusion was that the gravitational force of the early universe was stronger than present times and it has been diminishing continuously since then. This theory fell out of favor until George Gamow revived it in 1967, suggesting that instead a decrease in the gravitational force it should be a increase in the electrical force.

It seems now that as we know that expansion also generates an increase in volume of all matter,(Including protons, electrons as well as other particles and the distance between then) these ratios should keep also almost constant and if there is a decrease in gravitational force there should be also an equivalent decrease in the electrical force.

Other important numbers give by Rees are:

3) The number that measures the density of matter in the universe.

4) The number that indicates the force that keep the universe from fully expanding.

5) The amplitude of irregularities or ripples in the universe.

6) The number of spatial dimensions, which is 3.

All these numbers are derived after complicated equations which are considered far beyond the scope of this book but our point to bring this up is to check the precision of the first two numbers, a minimal change in the content of the primary components of our universe would result in complete failure to its development. This looks like close to explain God mathematically…

All these findings in different fields of science seem to indicate clearly a perfectly synchronized evolutive process from a cloud of loose particles to the complex, organized universe we are trying to understand this day.

The writer is not a sedimentologist, neither a mathematician or a theologist so he only prays that these ideas might help finding new answers from specialists on those fields.

# Chapter Eleven

## True And Relative Time

After dealing with all these arguments and theories the most important subject still left to be considered is time. St Augustine had the right answer when somebody asked him what happened before creation. His answer was that time started running together with creation as an integral part of it. We entirely agree with this statement, time was at the origin of everything and will keep on ticking unchanged till the last moment of the universe's existence.

This was the concept explained by Newton who assumed that there is one clock ticking equally for the entire universe. Contrary to this idea Einstein spoke of relative time and space denying the existence of absolute time for the universe and since then the concept of time have been bend, twisted, wrapped, stretched and reversed using all available mathematical formulae and the main conclusion is that it seems very clear that time is the main key of total control in the evolution of the universe.

Time's definition is close to God's, we know is there, we know that affect deeply our lives but we cannot see it or feel it and we cannot do a thing to change it, but the big difference is that time is God's creation and like anything else in the universe, it has limits. There was a time at the start of everything and there will be a time to end which will also be the end of time.

Einstein concluded there is an inseparable connection between time and the velocity of light and our observation is that light spreads evenly everywhere in certain fraction of time, but time itself is linear and moves in only one direction.

The scientist David Finkelstein developed a theory about time in which he described "chronos" as atoms of time. This theory was not very successful as it was not possible to have a reasonable explanation about the existence of chronos.The same happened to those theories that denied the value of time for being considered a man's creation.

The astrophysicist Thomas Gold proposed a theory that the arrow of time is directly related to the direction of heat flow out of the sun and in accordance to the second law of thermodynamics it would run from hot to cold, or from the sun to the outer planets of the system. He explained that should the universe be static, like a closed box, the arrow of time would be also static but considering the universe expansion, (and according to the principle that matter cools when expanded) the arrow of time will run outwards associated with the flow of energy out of the sun into the open space. But his conclusion was that there is a cyclic eternal universe with no beginning and no end, no Big Bang, no edge of the universe. He went even further to consider that if the universe were to contract everything would run backward, including the second law of thermodynamics and all other systems, including the human brain, so our memory would also run in reverse and we would remember the future…This is the case of a symmetric universe where everything grows in time and after reverses again to the beginning…????.

This concept of an eternal universe growing in cycles with no beginning and no end seems contrary to all the principles of nature, everything we know in our world, starting with our own Earth, its mountains and rivers and all its living flora, fauna and inhabitants have a time of birth, grow, maturity and death. Nothing is eternal but God…

Science has proved that everything else in the universe is changeable, reversible or recyclable, but real time is not, as it is the only measurement to show us where we come from, and where we are going. Time is the whole clue to solve the mystery of our existence, it started at the moment of creation and it kept running unchanged till today and will mark the end of everything when God decide is the end of the seventh day for our solar system.

Our descendants will see it coming, there will be marks in nature to show it but we will never know if and when will be an end for the whole universe.

A natural proof of the importance of time directly related to space in our world is shown by the ants, collecting food for winter at different times in different parts of the world, like bears fattening up for hibernation, farmers planning the next harvest, hunters the next season, fishermen the next moon tides, and NASA the next launching time. All together are directly related to Earth rotation, and the Moon rotating around us and our planet going around the sun.

*Ecclesiastes 3,*
*A time for everything.*
*There is a time for everything, and season for every activity*
*under heaven.*
*A time to be born and a time to die,*
*a time to plant and a time to uproot,*
*a time to kill and a time to heal,*
*a time to tear down and a time to build*
*a time to weep and a time to laugh,*
*a time to mourn and a time to dance,*
*a time to scatter stones and a time to gather them,*
*a time to embrace and a time to refrain,*
*a time to search and a time to give up/*
*a time to keep and a time to throw away,*
*a time to tear and a time to mend*
*a time to be silent and a time to speak,*

*a time to love and a time to hate,*
*a time for war and a time for peace,*

It seems that these verses of Ecclesiastes cover all aspects of creation, growth, evolution, maturity and the end of all times.

Later Einstein proved the same with his general theory of relativity, he said the space-time began with the Big Bang at the moment of creation and time would come to an end either at the big crunch singularity, if the whole universe recollapse, or compressed back to a simple blackhole. He thought that light is the only, and infalible way to measure the lenght and time of the universe.

So in essence if we ask…What is time?…Then the only answer is that time, like all the big laws of the universe, like music, like religion, like mathematics, is just there as another God's creation to be used by man to measure different events in sequence. The units used for these measurement are all man creation based on the movement of Earth around the sun. Therefore, time as we know it has only application inside the solar system. An inhabitant of another planet in the other end of our galaxy might have a day of 10 hours, a year of six month and might live an average of 180 years. Their measurement of the speed of light should be then something like 560,000 miles per second.

For us a second used to be defined as a 1/86,400 of a day, but man in his search to reach perfection proved that this is not an exact unit, (Another almost perfect act of God). Now one second is defined as 9,192,631,770 beats of a cesium atom, measured by the most precision clock on Earth, located in Bonn.

Then, as Paul Davis humorously quotes in his book About Time…"In this age of high precision time keeping poor old Earth does not make the grade"…

With all these, we must admit now that time is a result of our cultural conditions and as Einstein proved this earthly time is relative to our planet rotation around the sun but now if we try again to go back to the very instant when all started, when the true time was directly related to the birth of the whole universe, we might get some idea about absolute time.

According to Hawking space-time have no boundary meaning that it has no beginning, no moment of creation.

On the other hand, for some believers as well as for some pure scientists that believe in one single Big Bang, the universe is though to have had almost zero size and to have been almost infinitely hot at the very first moment of creation.

One second after Big Bang the temperature would have fallen to about ten thousand million degrees. (This is about the temperature at the center of the sun, like a H bomb explosion) and the universe would contain mainly photons, electrons and neutrinos equally distributed in this incredibly fast growing sphere that had a radius equal to 186,000 miles.(The time that light took to get to the edge) and from then on evolution took over to form atoms, molecules and matter that gave rise to the universe we now live in.

In his book "Afterglow of Creation, Marcus Chown describes the cosmic microwave background as the oldest fossil on creation, measured 15 billion years ago, when the Big Bang happened.

Since then evolution of matter took place until the formation of carbon, the key element for life. According to the American astrophysicis Robert Dicke, carbon was not there in the beginning, it was produced by super-nova explosions that keep recycling carbon into new generations of stars and he explains that it would take at least the time for one generation of stars to complete the cycle of birth, growing and death before biological life could get started. This theory could also fit with our idea of galaxies regeneration after reaching the point at critical speed near the edge of the universe.

Science is constantly discovering the laws that rule the universe and one of the latest, (originated while trying to justify further quantum mechanics), was the Uncertainty Principle, formulated by Heisenberg, who explained that one can never be exactly sure of the position and the velocity of a particle at a given time. If we project quantum mechanics to the cosmic world we will also have to project the uncertainty principle which might end bringing a philosophical principle of hopelessness about understanding the mechanics of the universe. In this sense it seems that Einstein was right in concluding the quantum mechanics is incomplete but, may be it is not if we manage to learn to use light as measurement.

The inclusion of time in physics seems to have turned the whole world more complicated then some scientists suggested to eliminate time in all equations but then, we have to consider it will be even worst. An example would be if we are at Denver airport waiting to get the 10.45 flight to Los Angeles, by eliminating time nobody will know whether we are in Denver, in the flying plane, or in Los Angeles.

Presently, there are some more, very elaborated new theories brought forward like Superstrings, these could help to find the answers we are seeking to understand the evolution of the universe till now and how it will develop in the near and far future.

Here we are trying work considering that time originated in a singularity, like Einstein and (in some stage) Hawkings believed it. This point is in the real center of the Big Bang, moving equally away along billions of radius lines running to the edge of the sphere of the universe, any measurement of this true radius must give the absolute time, meaning that, following the direction of the radius, time becomes irreversible, unchangeable and inflexible and if we pick a measure of true time along one of these radius and project it to all others, we will follow the circular surface of the sphere formed by the same measurement of true time on the whole universe.

Only when we measure time in other direction from the radius of the sphere formed by the universe time becomes relative.

Again we must remember that time has no mass, no charge, no energy, it moves at a constant speed in a direct line ending at the edge since the beginning. Any sphere inside will be younger than the whole creation, meaning that all stars in the universe will have different age if measured from the different spheres of equal time then, the older stars should be those closer to the edge, unless we are dealing with a slow rotating star that follows a more angular path from the radius.

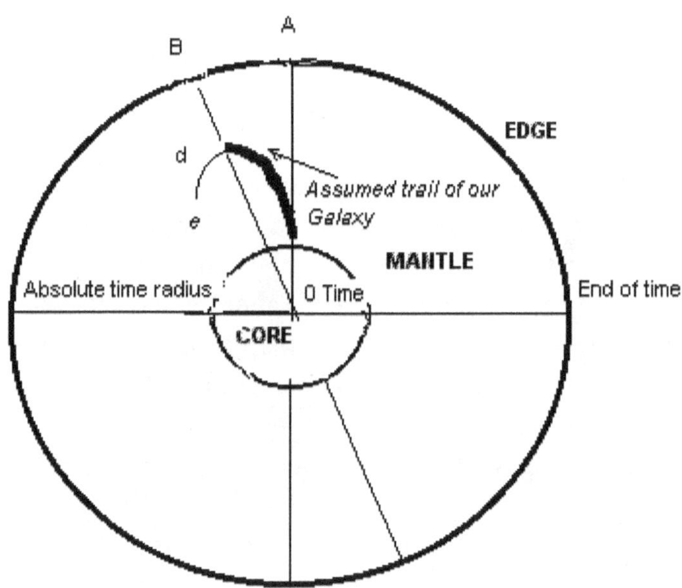


*Fig.11. Absolute and relative time. The absolute time is given by the radius of the sphere of the universe.*
*Relative time is measured from the Earth and the galaxies where we belong. We assume the path we follow might go from the Big Bang, along radius A, across the all absolute time lines to point d, somewhere along radius B. If we ever manage to travel back in time we might always end in an unknown zone like point e but never back where we were.*

Now if we consider our measurement of time from Earth which is waltzing with the Moon around the sun, and the whole Solar System that is moving around the Milky Way Galaxy, which is going around in circles with the other galaxies of the Local Group, then we have to agree that time as we measure it and in total agreement with Einstein principle, is relative, as all this circling about might indeed have taken us away from the direction of the radius of true time like is shown in figure 9.

This idea might also give room for the application of the Uncertainty Principle, as it seems very hard to think of a possibility for going back in time to the same space-time position where we were and where the Earth was in the solar system when the Milky Way with the whole galaxy group was traveling across radius A, as shown in figure 9.

In order to get back in time to the same space-time point we will have to trace back the moving of the Local Group, related to the Milky Way and to the sun around the galaxy and to the Earth and Moon dancing around the sun and then do all corrections for our Earth rotation so, if we ever manage to travel back in time it will be along the radius B, along the line of true time meaning that we might end nowhere into space…

John Gribbin has no doubt about time being affected by gravity and centrifugal forces and he thinks that time travelling might be possible in the proximity of a black hole using the strong gravitational field to go back in time as long as we want.

We agree that in all our complicated orbital and centrifugal planetary and stellar system, light might bend and time might seem to strech measured in "planetary units", but we know now that light is formed of particles which we understand have physical properties that could make them changeable, but time has no charge, no matter no energy, it's there and it cannot be changed, time measuring clocks change as there are man made material things but the absolute time as Newton thought, remains constant.

Quantum mechanics works in the assumption that, when conducting an experiment, as soon as you try to measure the movement of a particle you disturb all the system failing to find the position of such particle at certain time due to the Uncertanty Principle. This is clear evidence that time itself is what creates the uncertainty and following Einstein's conviction, light might be the only instrument to make an accurate measurement of time.

In conclusion we wish to stress the importance of understanding the difference between the days of God's creation and the 24 hour days picked by men from nature. There are two kinds of time, one is God's time irreversible and always adding, never static, never declining, controlling the growth of the whole universe. The other time is man creation, man imitation of God to measure his own growth in the world he lives in.

The difference again is that all our measurements are relative, like Einstein experiment when dropping a stone from a moving train. The person that drops it, see the stone falling straight down, but the man watching from the embankment see the parabola the stone draws while falling. Absolute time does not bend, our measuring stone does.

The idea of two time scales is not new, Arthur Milner, who held the Chair of Mathematics at Oxford,deduced that there are two time scales of significance which he designated with Greek letters to define the time of atomic processes related to the speed of light for one and the time to measure gravitational forces and other main processes like Earth rotation and other movements related to the sun. He defended the idea that atomic clocks gradually will march ahead of astronomical clocks that measure the lenght of our days and years.

The differences measures with this theory is very small to be considered but, going back closer to Big Bang times the difference mounts up and measured in Earth's time the universe would become infinitely old.

Contrary to conventional physics he also believed that the laws of physics ought to follow from the nature of the universe and it would be possible to find the main laws by following all the process from the first clouds of matter, through expansion and the formation of stars and planets. Contrary to this, most of today scientists think that cosmologists have to use the existent laws to explain the universe

The famous British physicist Paul Dirac also suggested the idea of two times during the 1930's. He also thought that atomic clocks moved ahead of astronomical clocks as the planets move away from each other due to expansion and enlargement of their orbits and according to Dirac this movements would cause a variation in the strength of the force of gravity of the planets in the solar system.

The landing of the Viking in Mars allowed to take accurate measurements of time and distances across the solar system and permitted to determine that as the planets movement are constantly affecting each other, making almost impossible to make any measurement in changes of gravity then, after years of working in the data obtained on this mission, this Dirac's theory was also dropped.

Going back to Thomas Gold theory about time following the direction of the flow of energy from hot to cold, if we apply this idea to our model of evolution of the universe, we could also establish the existence of two sources of heat, the heat generated from the sun that result in our solar system time scale and the heat generated after the Big Bang that gives the scale of the universal time, coming from the unknown center of the universe.

Concerning the constancy of light, Einstein found that whenever we make a measurement of the velocity of light, regardless of whether we are in motion or at rest relative to the light source, we always get the same result; 186,000 miles per second. The bible said that God's first words on creation were *"Let there we light". Then* light was on from the very first instant of creation then there must be a connection between space-time and light. Light covers all space at the same time throughout the whole universe and can only be measured along the line of true time from the center to the edge of the universe.

*Psalm 102…: 25. In the beginning you laid the foundations of the earth, and the heavens are the work of your hands.*
*They will perish, but you remain; they will wear out like a garment.*
*Like clothing you will change them and they will be discarded.*
*But you remain the same, and your years will never end.*
*The children of your servants will live in your presence; their descendants will be established before you."*

On the writer understanding, quantum mechanics, like all the following latest theories are a mixture of elegant mathematics applied to philosophical concepts about subatomic particles. From all these our modern technology has benefited greatly for their findings while working in the subatomic world but there is still a big gap to project these concepts to the macro world of the cosmology. The use of superstrings seem to have achieve some results but in some cases have led to some bizarre conclusions like the one about existence of many universes, or of mini universes forming all around us.

All in all, with a little bit of refining we might be getting closer to the final touches of science that will eventually bring us to find an explanation about all the laws of the universe and then we will finaly succeed in proving the existence of God mathematically as the well known mathematician Kurt Godel already tried without being taken seriously as he also talked about time travelling in an infinite, static universe.

We first have to accept the bible as the word of God and the carrier of undeniable information on the world's and maybe the whole universe creation, then we can study the book of Genesis with a purely scientifical approach and might be able to trim further the length of each day of creation and from there, work out the measurement of time used by God. Is the bible talking about creation of our Earth alone…? or maybe our Solar System…? or the Galaxy…?

This book of the bible talks only about the world's creation. It never mentions the universe…

We only know the relative time using Solar System units, and even if scientists use atomic clocks to have more accurate measurements we are still measuring time using solar system components then, Newton may be still right, as we might still need some unknown universal and absolute time measurement units to determine the existence of our universe from the alpha to the omega…

In the first six chapters we tried to explain the gigantic work that took our Creator to form the planet we live and all the modification he did while improving to make it comfortable for us. All this tremendous effort should make us feel proud of the importance we have on God's eyes, however, we know now that our Earth, being a major planet in our Solar System still is a minor dot compared to the immensity of the Milky Way Galaxy, which is itself only a part of the Local Group of Galaxies, and altogether are a small spot in the vast universe.

Knowing all this, even the most devout religious theologist should admit that we would be presumptuous in thinking that all this was made only for our benefit…

# Bibliography


Anderson, Roger.—Marine Geology. John Wiley and Sons.New York.1986

Arbor, Ann. More Evidence for the Impact Origin of the Moon. University of Michigan.1997

Barbour, Julian The End of Time. Oxford Univ.Press.2000

Barrow, John D and Silk, Joseph. The Left Hand of Creation. Basic. Books,Inc.New York.1983

Cameron, A.G.W. Cameron.-The Origin of the Moon and the Single Impact Hypothesis. Harvard College. Cambridge.1999

Chown, Marcus.-Afterglow of creation.

Davies, Paul. About Time. Published By Simon and Shuster. New York. 1995

Davies, Paul.-The Mind of God.Publ. by Simon &Schuster. New York.1992.

Erickson, Jon. Marine Geology.Facts of File. New York.1996.

Erikson.Jon. The Ice Ages

Ferguson, K. Stephen Hawking-.Quest for a Theory of Everything. a Bantam Book.NY. 1987

Ferris, Timothy. The Whole Shebang.Simon & Schuster. New York. 1997

Frommert, H and Kronberg, C.-Yahoo's index of Milky Way web-pages. April 1999

Goldsmith D and Cohen N, Mysteries of the Milky Way. Contemporary Books. Chicago 1991

Gribbin, John-In The Beginning. Little, Brown and Company. New York.1993

Gribbin, John.-In the Search of the Edge of Time. Penguin Books. London.1992

Gribbin, John,-When Time Began. Yale Univ.Press. New Haven 1999.

Hawkins, Stephen.—A Brief History of Time. Bantam Book. 1998

HOLY BIBLE.-New International Version.Bible Society of South Africa. 1978

Kaku, Michio and Thompson J.,-Beyond Einstein. Anchor Books. New York1995

MacLeod. Norman.-Mass Extinction at the Cretaceous Tertiary bounday. Dep. of Paleontology. The Natural History Museum. London. 1999.

Moore Patrick—Exploring the Earth and Moon. Gallery Books. New York.1991

Sklar, Lawrence.-Space,Time, and Spacetime.Univ.of California Press. Los Angeles 1974

Smoot,G and Davidson K.-Wrinkles in Time. Avon Books. New York 1993.

Spudis, Paul. D, The Once and Future Moon. Smithsonian Institute Press. 1996

St. Augustine.-City of God. Translated by M.Dods. The Modern Library New York. 1950

Towson, Earl.-Origin of the Moon. www.astronomy.palomar.edu.

Trefil, James.-Reading the mind of God. Charles Scrobner and Sons. New York. 1989.

Tucker, W and K, The Dark Matter.W.Morrow&Co.,Inc. New York. 1988.

Weinberg, Stephen.-The First Three Minutes

Wilhelms, Don E.-To a Rocky Moon.The University of Arizona Press. 1993

www.ucl.ac.uk.-The Origin of the Moon. Internet 1999.

www.permanent.com.-The origin and Composition of the Moon. 1977

Zukav, Gary.-The Dancing Wu Li Masters.Bantam New Age Books. New York, 1980

Click here to input the text of your bibliography, if any. Entries should be listed alphabetically...each entry occurring on a separate line.)

www.ingramcontent.com/pod-product-compliance
Lightning Source LLC
Chambersburg PA
CBHW031259060726
47590CB00003B/981